OLIGONUCLEOTIDE THERAPEUTICS

First Annual Meeting of the Oligonucleotide Therapeutics Society

ANNALS OF THE NEW YORK ACADEMY OF SCIENCES
Volume 1082

OLIGONUCLEOTIDE THERAPEUTICS

First Annual Meeting of the Oligonucleotide Therapeutics Society

Edited by Thomas Tuschl and John Rossi

Published by Blackwell Publishing on behalf of the New York Academy of Sciences
Boston, Massachusetts
2006

Library of Congress Cataloging-in-Publication Data

Oligonucleotide Therapeutics Society. Meeting (1st : 2005 :
New York, N.Y.)
Oligonucleotide therapeutics / First Annual Meeting of the
Oligonucleotide Therapeutics Society ; edited by Thomas Tuschl
and John Rossi.
p. ; cm. – (Annals of the New York Academy of Sciences,
ISSN 0077-8923 ; v. 1082)
"This volume is the result of a conference entitled First Meeting of
the Oligonucleotide Therapeutics Society held by the New York
Academy of Sciences and the Oligonucleotide Therapeutics Society
on September 15–18, 2005 at Rockefeller University in New York
City."
Includes bibliographical references and index.
ISBN-13: 978-1-57331-587-6 (alk. paper)
ISBN-10: 1-57331-587-7 (alk. paper)
1. Antisense RNA–Therapeutic use–Congresses. I. Tuschl, Thomas.
II. Rossi, John J. III. New York Academy of Sciences. IV. Title.
V. Series.
[DNLM: 1. Oligonucleotides–Therapeutic use–Congresses.
2. Oligonucleotides–therapeutic use–Congresses.
3. Oligonucleotides–pharmacokinetics–Congresses.
W1 AN626YL v.1082 2006 / QU 57 O46o 2006]
RM666.N87O448 2006
615'.31–dc22

2006024996

The *Annals of the New York Academy of Sciences* (ISSN: 0077-8923 [print]; ISSN: 1749-6632 [online]) is published 28 times a year on behalf of the New York Academy of Sciences by Blackwell Publishing, with offices located at 350 Main Street, Malden, Massachusetts 02148 USA, PO Box 1354, Garsington Road, Oxford OX4 2DQ UK, and PO Box 378 Carlton South, 3053 Victoria Australia.

Information for subscribers: Subscription prices for 2006 are: Premium Institutional: $3850.00 (US) and £2139.00 (Europe and Rest of World).
Customers in the UK should add VAT at 5%. Customers in the EU should also add VAT at 5% or provide a VAT registration number or evidence of entitlement to exemption. Customers in Canada should add 7% GST or provide evidence of entitlement to exemption. The Premium Institutional price also includes online access to full-text articles from 1997 to present, where available. For other pricing options or more information about online access to Blackwell Publishing journals, including access information and terms and conditions, please visit www.blackwellpublishing.com/nyas.

Membership information: Members may order copies of the *Annals* volumes directly from the Academy by visiting www.nyas.org/annals, emailing membership@nyas.org, faxing 212-888-2894, or calling 800-843-6927 (US only), or +1 212 838 0230, ext. 345 (International). For more information on becoming a member of the New York Academy of Sciences, please visit www.nyas.org/membership.

Journal Customer Services: For ordering information, claims, and any inquiry concerning your institutional subscription, please contact your nearest office:
UK: Email: customerservices@blackwellpublishing.com; Tel: +44 (0) 1865 778315;
Fax +44 (0) 1865 471775
US: Email: customerservices@blackwellpublishing.com; Tel: +1 781 388 8599 or
1 800 835 6770 (Toll free in the USA); Fax: +1 781 388 8232
Asia: Email: customerservices@blackwellpublishing.com; Tel: +65 6511 8000;
Fax: +61 3 8359 1120
Members: Claims and inquiries on member orders should be directed to the Academy at email: membership@nyas.org or Tel: +1 212 838 0230 (International) or 800-843-6927 (US only).

Printed in the USA.
Printed on acid-free paper.

Mailing: The *Annals of the New York Academy of Sciences* are mailed Standard Rate. **Postmaster:** Send all address changes to *Annals of the New York Academy of Sciences*, Blackwell Publishing, Inc., Journals Subscription Department, 350 Main Street, Malden, MA 01248-5020. Mailing to rest of world by DHL Smart and Global Mail.

Annals are available to subscribers online at the New York Academy of Sciences and also at Blackwell Synergy. Visit www.annalsnyas.org or www.blackwell-synergy.com to search the articles and register for table of contents e-mail alerts. Access to full text and PDF downloads of *Annals* articles are available to nonmembers and subscribers on a pay-per-view basis at www.annalsnyas.org.

The paper used in this publication meets the minimum requirements of the National Standard for Information Sciences Permanence of Paper for Printed Library Materials, ANSI Z39.48_1984.

ISSN: 0077-8923 (print); 1749-6632 (online)
ISBN-10: 1-57331-587-7 (alk. paper); ISBN-13: 978-1-57331-587-6 (alk. paper)

A catalogue record for this title is available from the British Library.

Digitization of the *Annals of the New York Academy of Sciences*

An agreement has recently been reached between Blackwell Publishing and the New York Academy of Sciences to digitize the entire run of the *Annals of the New York Academy of Sciences* back to volume one.

The back files, which have been defined as all of those issues published before 1997, will be sold to libraries as part of Blackwell Publishing's Legacy Sales Program and hosted on the Blackwell Synergy website.

Copyright of all material will remain with the rights holder. Contributors: Please contact Blackwell Publishing if you do not wish an article or picture from the *Annals of the New York Academy of Sciences* to be included in this digitization project.

ANNALS OF THE NEW YORK ACADEMY OF SCIENCES

Volume 1082
October 2006

OLIGONUCLEOTIDE THERAPEUTICS

First Annual Meeting of the Oligonucleotide Therapeutics Society

Editors
THOMAS TUSCHL AND JOHN ROSSI

This volume is the result of a conference entitled **First Meeting of the Oligonucleotide Therapeutics Society** held by the New York Academy of Sciences and the Oligonucleotide Therapeutics Society on September 15–18, 2005 at Rockefeller University in New York City.

CONTENTS

Preface and Overview. *By* THOMAS TUSCHL AND JOHN ROSSI xi

Part I. Delivery

Surface-Modified LPD Nanoparticles for Tumor Targeting. *By* SHYH-DAR LI AND LEAF HUANG 1

Atelocollagen-Mediated Systemic DDS for Nucleic Acid Medicines. *By* KOJI HANAI, FUMITAKA TAKESHITA, KIMI HONMA, SHUNJI NAGAHARA, MIHO MAEDA, YOSHIKO MINAKUCHI, AKIHIKO SANO, AND TAKAHIRO OCHIYA 9

Intracellular Delivery of Oligonucleotide Conjugates and Dendrimer Complexes. *By* R.L. JULIANO 18

Mechanism of PNA Transport to the Nuclear Compartment. *By* LENKA STANKOVA, AMY J. ZIEMBA, ZHANNA V. ZHILINA, AND SCOT W. EBBINGHAUS 27

Part II. Identifying and Validating Targets and Systems

TLR9 and the Recognition of Self and Non-Self Nucleic Acids. *By* MARC S. LAMPHIER, CHERILYN M. SIROIS, ANJALI VERMA, DOUGLAS T. GOLENBOCK, AND EICKE LATZ 31

RNA Silencing in the Struggle against Disease. *By* VOLKER PATZEL, ISABELL DIETRICH, AND STEFAN H.E. KAUFMANN 44

Cellular Dynamics of Antisense Oligonucleotides and Short Interfering RNAs. *By* LI KIM LEE, BRANDEE M. DUNHAM, ZHUTING LI, AND CHARLES M. ROTH 47

Inhibition of Hepatitis C IRES-Mediated Gene Expression by Small Hairpin RNAs in Human Hepatocytes and Mice. *By* HEINI ILVES, ROGER L. KASPAR, QIAN WANG, ATTILA A. SEYHAN, ALEXANDER V. VLASSOV, CHRISTOPHER H. CONTAG, DEVIN LEAKE, AND BRIAN H. JOHNSTON 52

SiRNA-Mediated Selective Inhibition of Mutant Keratin mRNAs Responsible for the Skin Disorder Pachyonychia Congenita. *By* ROBYN P. HICKERSON, FRANCES J.D. SMITH, W.H. IRWIN MCLEAN, MARKUS LANDTHALER, RUDOLF E. LEUBE, AND ROGER L. KASPAR 56

Multitargeted Approach Using Antisense Oligonucleotides for the Treatment of Asthma. *By* Z. ALLAKHVERDI, M. ALLAM, A. GUIMOND, N. FERRARI, K. ZEMZOUMI, R. SÉGUIN, L. PAQUET, AND P.M. RENZI 62

Therapeutic Modulation of *DMD* Splicing by Blocking Exonic Splicing Enhancer Sites with Antisense Oligonucleotides. *By* A. AARTSMA-RUS, A.A.M. JANSON, J.A. HEEMSKERK, C.L. DE WINTER, G.-J.B. VAN OMMEN, AND J.C.T. VAN DEUTEKOM 74

Neuromuscular Therapeutics by RNA-Targeted Suppression of ACHE Gene Expression. *By* AMIR DORI AND HERMONA SOREQ 77

Characterization of Antisense Oligonucleotides Comprising 2′-Deoxy-2′-Fluoro-β-D-Arabinonucleic Acid (FANA): Specificity, Potency, and Duration of Activity. *By* NICOLAY FERRARI, DENIS BERGERON, ANNA-LISA TEDESCHI, MARIA M. MANGOS, LUC PAQUET, PAOLO M. RENZI, AND MASAD J. DAMHA 91

Anti-HIV Activity of Steric Block Oligonucleotides. *By* GABRIELA IVANOVA, ANDREY A. ARZUMANOV, JOHN J. TURNER, SANDRINE REIGADAS, JEAN-JACQUES TOULMÉ, DOUGLAS E. BROWN, ANDREW M.L. LEVER, AND MICHAEL J. GAIT 103

Selection of Thioaptamers for Diagnostics and Therapeutics. *By* XIANBIN YANG, HE WANG, DAVID W.C. BEASLEY, DAVID E. VOLK, XU ZHAO, BRUCE A. LUXON, LEE O. LOMAS, NORBERT K. HERZOG, JUDITH F. ARONSON, ALAN D.T. BARRETT, JAMES F. LEARY, AND DAVID G. GORENSTEIN 116

Modification of the Pig CFTR Gene Mediated by Small Fragment Homologous Replacement. *By* ROSALIE MAURISSE, JUDY CHEUNG, JONATHAN WIDDICOMBE, AND DIETER C. GRUENERT 120

Part III. Clinical Studies

Nucleic Acid Therapeutics for Hematologic Malignancies—Theoretical Considerations. *By* JOANNA B. OPALINSKA, ANNA KALOTA, JYOTI CHATTOPADHYAYA, MASAD DAMHA, AND ALAN M. GEWIRTZ 124

CpG Oligonucleotides Improve the Protective Immune Response Induced by the Licensed Anthrax Vaccine. *By* DENNIS M. KLINMAN, HANG XIE, AND BRUCE E. IVINS 137

Anti-VEGF Aptamer (Pegaptanib) Therapy for Ocular Vascular Diseases. *By* EUGENE W.M. NG AND ANTHONY P. ADAMIS 151

RNAi in Combination with a Ribozyme and TAR Decoy for Treatment of HIV Infection in Hematopoietic Cell Gene Therapy. *By* MINGJIE LI, HAITANG LI, AND JOHN J. ROSSI 172

Index of Contributors 181

Financial assistance was received from:

Silver Sponsors

- Avecia
- Coley Pharmaceutical Group, Inc.
- Isis Pharmaceuticals
- Pfizer, Inc.
- National Institutes of Health/National Cancer Institute

Bronze Sponsors

- Alnylam Pharmaceuticals
- ALTANA Pharma AG
- Avatar Pharmaceutical Services Inc.
- Calando Pharmaceuticals
- Dharmacon Inc.
- Dowpharma
- Dynavax Technologies
- Ercole Biotech Inc.
- Eurogentec
- Eyetech Pharmaceuticals
- GE Healthcare
- Genta
- Glen Research Corporation
- IDT
- Invitrogen
- Kinovate Life Sciences, Inc.
- Merck
- Monomer Sciences Inc.
- Pierce Nucleic Acid Technologies
- PrimeSyn Lab Inc.
- Promega
- Sigma-Aldrich Corp.
- Topigen

Preface and Overview

The gene-silencing techniques that show so much promise for therapeutic strategies and related basic research in functional genomics are based on our understanding of the actions of oligonucleotides. This volume presents articles describing research presented at the first annual meeting of the Oligonucleotide Therapeutics Society, sponsored by the New York Academy of Sciences and was held at The Rockefeller University in New York City in September 2005. It was attended by over 250 scientists interested in the biology and applications of oligonucleotides. General lectures on RNA processing and metabolism and RNAi mechanisms were presented on the first day. These lectures were followed by a series of specialized lectures covering the entire breadth of oligonucleotide applications. Research topics covered during the meeting included new oligonucleotide chemistries, antisense oligonucleotide therapy, the biology and applications of small interfering RNAs, aptamer technology, oligonucleotide-directed alternative splicing, triplex-forming oligonucleotides, immunostimulatory oligonucleotide biology and applications, and oligonucleotide delivery strategies. The meeting was attended by scientists from academic institutions as well as from industry, with a special session devoted to the industrial sector of the Oligonucleotide Therapeutics Society.

This volume provides a smorgasbord of oligonucleotide-based studies and applications and inaugurates what we hope will become an annual publication. The organization of the reports included here takes a slightly different form than the conference program and begins with a section on oligonucleotide delivery strategies. Li and Huang report on a liposome–polycation–DNA (LPD) complex for short interfering RNA (siRNA). Surface modification of the LPD complex increased delivery efficiency and gene silencing of survivin in human lung cancer cells. An atelocollagen–oligonucleotide complex of nano-sized particles developed by Hanai *et al.* successfully delivered siRNA to metastasized tumors in bone tissue and suppressed ear dermatitis in a contact hypersensitivity model.

The approach Juliano took for enhancing delivery of antisense and siRNA molecules involved using cell-penetrating peptides (CPPs). These soluble conjugates may have advantages over particle-based delivery *in vivo*. Stankova *et al.* evaluated nuclear uptake of fluorescently labeled peptide nucleic acids (PNAs) and present their conclusion that PNAs are transported into the nucleus through an energy-dependent process involving the nuclear pore complex.

Eicke Latz's group (Lamphier *et al.*) looked at neuroimmunomodulation through the actions of toll-like receptors (TLRs) in response to invading pathogens. They report that, because TLR9 binds to a broad range of DNAs, an

Ann. N.Y. Acad. Sci. 1082: xi–xiii (2006). © 2006 New York Academy of Sciences.
doi: 10.1196/annals.1348.067

additional recognition step is likely to be involved in its sensing of differences in the structures of bound DNA.

The article by Patzel, Dietrich, and Kaufmann is included in the section of the volume that covers the identification and validation of targets and systems with an article on RNA silencing. Their techniques allowed them to trigger gene silencing by siRNA in prokaryotic cells, which previously has required the application of knock-out or antisense techniques, both of which are less efficient than siRNA. Later in the volume, Li, Li, and Rossi investigated the potential of gene therapy for the treatment of HIV infection. Because the use of a single gene as the inhibitory agent is not practical for long-term therapy, they have combined various HIV-based inhibitors into a lentiviral vector for transduction of hematopoietic progenitor cells, which can then be reinfused into HIV patients. They used three antiviral genes, including a short hairpin RNA targeting an HIV tat/rev common exon, a nucleolar-localizing TAR decoy, and a ribozyme that degrades the CCR5 mRNA. Their article discusses the efficacy of this combination as a treatment for HIV infection.

The dynamics of the effectiveness of antisense oligonucleotides (AS ODNs) and siRNAs were compared by Lee *et al.*, who report that certain regions of mRNA may be susceptible to both AS ODNs and siRNAs. Kaspar *et al.* looked at the ability of small hairpin RNAs (shRNAs) to inhibit hepatitis C internal ribosome entry site-dependent gene expression. Their results indicate that shRNAs have potential for controlling the hepatitis C virus.

The promising potential of the use of RNAi for treating genetic disorders is reported on in the article contributed by Hickerson *et al.*, who used the rare monogenic skin disorder, pachyonychia congenital, as a model. Allakhverdi *et al.* discuss the potential for using two oligonucleotides simultaneously to prevent eosinophilia and airway hyperresponsiveness in asthma, and address the challenges inherent in the development of a product using two oligonucleotides in humans. The use of AS ODN to correct the disrupted reading frame of Duchenne muscular dystrophy patients is the focus of the article by Aartsma-Rus *et al.*, and the following article, by Dori and Soreq, presents the use of Monarsen, a 20-mer acetylcholinesterase-targeted antisense agent, in the treatment of myasthenia gravis. Ferrari *et al.* present the development of phosphorothioate antisense deoxyribonucleotides (PS-DNA), which may be useful in a variety of therapeutic applications.

Another approach to combating HIV infection is presented by Ivanova *et al.* in an article describing the use of oligonucleotide analogues that act as steric block agents of the HIV RNA function. In their lab, they have targeted the HIV-1 *trans*-activation responsive region and the viral packaging signal with steric block oligonucleotides and demonstrated their potential for steric blocking of viral protein interactions both *in vitro* and in cells. Their first antiviral studies are described here. Thioaptamers show enhanced affinity, specificity, and higher stability compared to normal phosphate ester backbone aptamers, and Yang *et al.* describe their development of *in vitro* thioaptamer selection and bead-based thioaptamer selection techniques and identification of thioaptamers targeting specific proteins, such as transcription factor NF-κB and AP-1 proteins. In the final article of the section on identifying and validating targets and systems, Maurisse *et al.* have written about their development of a pig model of cystic fibrosis. They chose the pig as their candidate for a large-animal model for cystic fibrosis because of its anatomical and physiological similarities to the human, particularly the airways.

Clinical studies reported in this volume include a look at the nucleic acid therapeutics of hematologic malignancies by Opalinska *et al.*, including some discussion of the disappointing outcomes of clinical trials of antisense nucleic acid drugs. Klinman, Xie, and Ivins describe combining synthetic oligodeoxynucleotides containing unmethylated CpG motifs with Anthrax Vaccine Adsorbed (the licensed human vaccine) and observing increases in speed, magnitude, and avidity of the antibody response. The volume's penaltimate article is by Ng and Adamis, in which they present their work with vascular endothelial growth factor (VEGF), which regulates physiological and pathological angiogenesis, and pegaptanib, which specifically inhibits the $VEGF_{165}$ isoform and shows potential for treating two diseases associated with ocular neovascularization: age-related macular degeneration and diabetic macular edema.

The editors would like to express their appreciation to all of those who contributed to the success of this meeting and to its published record in this volume, beginning with our sponsors, who are listed in the Table of Contents. The meeting would not have been possible without the efforts of our colleagues on the organizing committee: Fritz Eckstein, Alan M. Gewirtz, Gunther Hartmann, Ryszard Kole, Art Krieg, Bernard LeBleu, Brett Monia, Georg Sczakiel, and Cy Stein. We would also like to thank Rashid Shaikh, Shari Dermer, and Renee Wilkerson-Brown of the New York Academy of Sciences for their guidance and logistical support. The staff of the editorial department of the *Annals* has also ensured the timely publication of this volume. We particularly thank Kirk Jensen, Steve Bohall, Hilary Burdge, and Linda Mehta, whose combined efforts guided this volume through to press.

THOMAS TUSCHL
The Rockefeller University
New York, New York

JOHN ROSSI
Beckman Research Institute, City of Hope
Duarte, California

Surface-Modified LPD Nanoparticles for Tumor Targeting

SHYH-DAR LI AND LEAF HUANG

Division of Molecular Pharmaceutics, School of Pharmacy, University of North Carolina at Chapel Hill, Chapel Hill, North Carolina 27599, USA

ABSTRACT: We have developed a tumor-targeted LPD formulation (liposome-polycation-DNA complex) for siRNA. With surface modification, the targeted, PEGylated LPD increased the delivery efficiency by four-fold and the gene-silencing effect by two- to three-fold. Downregulation of survivin in human lung cancer cells by targeted LPD induced 90% of apoptosis and sensitized the cells to cisplatin by four-fold. PEGylated LPD formulation also significantly improved the tumor localization of siRNA in the NCI-H460 human lung cancer xenograft model. The tumor appeared to be the major uptake organ for siRNA formulated in surface-modified LPD. Our encouraging results indicate that surface-modified LPD may be a potent carrier for RNAi-based tumor therapy.

KEYWORDS: systemic delivery; siRNA; LPD nanoparticles; liposome; PEGylation; tumor targeting

INTRODUCTION

Ever since siRNA was shown to silence both exogenous and endogenous genes in mammalian cells,[1] there has been increasing enthusiasm for developing therapies based on RNA interference. However, siRNA delivery via systemic routes remains a challenge due to its negatively charged backbone and the instability in physiological fluids.

We have reviewed the last 2 years of progress on the development of systemically used delivery systems for siRNA.[2] We concluded that PEGylated nanoparticles with multiple components are the most promising nonviral vectors for systemic delivery. Cationic molecules, such as cationic lipids[3] and polyethyleneimine (PEI),[4] which can interact with negatively charged siRNA, are the major components of the delivery systems. To prevent the aggregation of the resulting complex with serum components, surface steric hindrance is introduced by PEGylation. Thus, PEGylated nanoparticles (~100 nm) accumulate in highly vasculated tissues, referred to as the EPR (enhanced permeability and

Address for correspondence: Leaf Huang, Ph.D., CB 7360, 2316 Kerr Hall, Chapel Hill, NC 27599-7360. Voice: 919-843-0736; fax: 919-966-0197.
e-mail: leafh@unc.edu

Ann. N.Y. Acad. Sci. 1082: 1–8 (2006).
doi: 10.1196/annals.1348.001

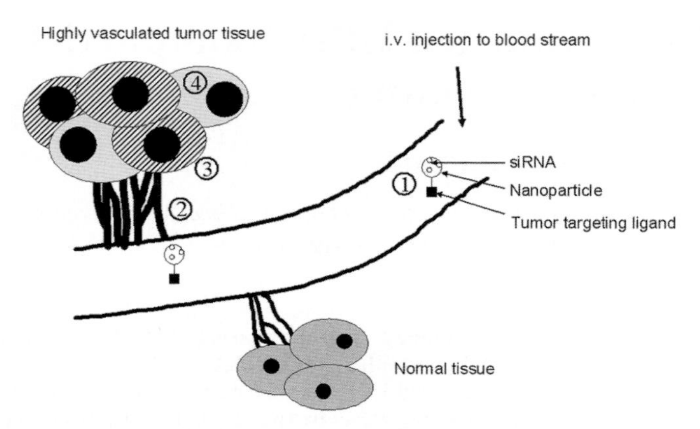

FIGURE 1. Barriers of tumor-targeted delivery and multicomponent design of a siRNA carrier. (1) stability in the blood; (2) extravasation and accumulation in tumors; (3) intracellular uptake by tumor cells and (4) siRNA release in the cytosol of cells.

retention) effect.[5] Ligands are attached to the end of PEG to increase cellular bioavailability.[6] Endosomolytic agents, such as PEI[4] and fusogenic lipids,[3] are incorporated in the formulation to facilitate intracellular release of siRNA. The strategies for designing a tumor-targeted delivery vector are summarized in FIGURE 1.

Although significant progress has been made over the last 2 years, there are still some issues that need to be addressed, such as high liver uptake and relatively low tumor accumulation of the vectors.

In this article, we discuss our formulation work and *in vitro* experiments showing the efficient delivery and gene-silencing effects achieved by the targeted LPD. Preliminary *in vivo* data of tissue distribution demonstrated the potential of the formulation as an siRNA carrier for the solid tumor.

LPD AND PEGYLATED LPD PREPARATION

The LPD and PEGylated LPD nanoparticles were prepared as previously described with minor modification (FIG. 2).[7] Briefly, small unilamellar liposomes consisting of DOTAP and cholesterol (1:1 molar ratio) were prepared by thin film hydration followed by membrane extrusion. The total lipid concentration of the liposome was fixed at 10 mM. LPD comprised DOTAP/cholesterol liposomes, protamine, and a mixture of siRNA and calf thymus DNA (1:1 weight ratio). To prepare LPD, 18 μL protamine (2 mg/mL), 138 μL deionized water, and a 24-μL mixture of siRNA and calf thymus DNA (2 mg/mL) were mixed in a 1.5 mL tube. The complex was allowed to stand at room temperature for 10 min before the addition of 120 μL of DOTAP/cholesterol liposomes (total lipid concentration = 10 mM). LPD nanoparticles were kept at room

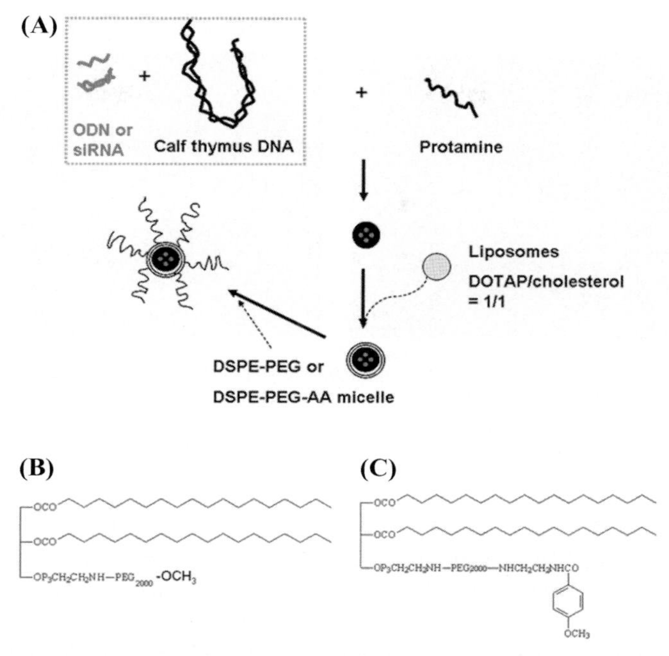

FIGURE 2. Illustration of preparation of PEGylated LPD (**A**) and chemical structures of DSPE-PEG_{2000} (**B**) and DSPE-PEG_{2000}-anisamide (**C**). (Reproduced with permission from Li and Huang, 2006.[7])

temperature for another 10 min before further application. PEGylated LPD formulations were prepared by incubating 300 μL preformed LPD with 15–30 μL of micelle suspension of DSPE-PEG or DSPE-PEG-anisamide (25 mg/mL) and then incubated at 50°C for 10 min. Anisamide is a small molecule ligand that targets to the sigma receptor overexpressing cells, e.g., human lung cancer cells. The resulting formulations were allowed to cool to room temperature before use. The encapsulation efficiency of siRNA in LPD formulations was determined by passing the fluorescein-labeled siRNA (FAM-siRNA) containing formulations through a Sepharose CL4B size exclusion column (Pharmacia Biotech, Uppsala, Sweden). Unencapsulated FAM-siRNA was collected and quantified and incorporation efficiency was calculated.

In our formulation, a carrier DNA, calf thymus DNA was employed to encapsulate siRNA in the condensed core with the help of protamine. The complex was then coated with cationic liposomes and assembled into uniformly sized nanoparticles. The formulation was optimized by fine-tuning the ratio of siRNA/protamine/liposomes and different compositions of the liposomes. Postinsertion of DSPE-PEG into the lipid bilayers of LPD was shown to shield the positive charge, bring the nanoparticles to neutral, and prevent aggregation

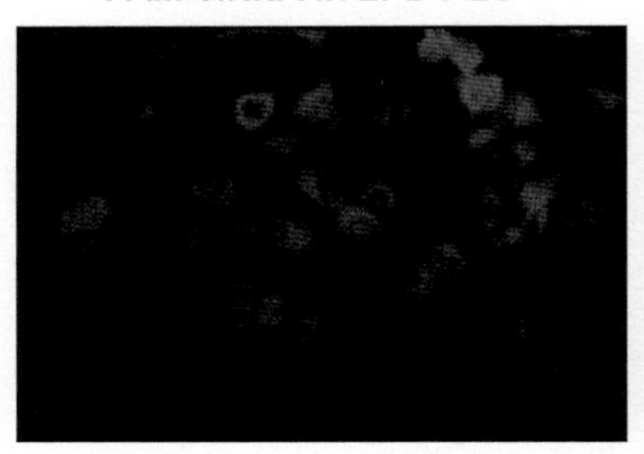

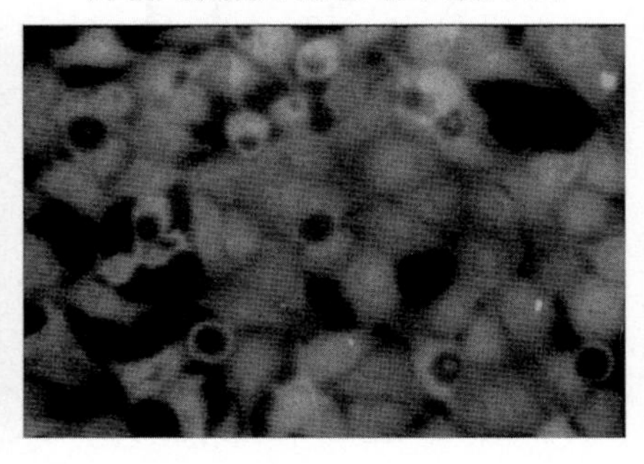

FIGURE 3. Fluorescence photographs of NCI-H1299 cells after treatment with 5′ FAM-labeled siRNA in LPD-PEG or LPD-PEG-AA. Cells were treated with different formulations at 37°C for 4 h. Cells were rinsed, fixed, and imaged by a fluorescence microscope. Magnification = 400×. (Reproduced with permission from Li and Huang, 2006.[7])

in the serum.[7] The more the DSPE-PEG was introduced, the better the shielding was. For our LPD formulation, 20–25 mol% of DSPE-PEG input almost completely shielded the surface charge of LPD nanoparticles. Fewer amounts of PEG lipids with higher molecular weight were required to achieve the same effect (data not shown). The zeta potential, particle size, and incorporation efficiency of the optimized, surface-modified LPD were 4.27 mV, 120–150 nm, and >95%, respectively.

CELLULAR UPTAKE AND GENE-SILENCING STUDY

As shown in FIGURE 3, the fluorescence signal in the cells treated with the targeted formulation, LPD-PEG-anisamide (LPD-PEG-AA), was significantly stronger than that of cells treated with the nontargeted formulation, LPD-PEG. It indicates that the anisamide ligand increased the delivery efficiency of the nanoparticles for sigma receptor-expressing cells, NCI-1299. Our quantitative data showed that the delivery efficiency of LPD-PEG-AA was four-fold higher compared to nontargeted formulation and could be partially competed by excess free haloperidol, which is a known ligand for sigma receptors. These results suggest that LPD-PEG-AA targeted to NCI-1299 cells via a sigma receptor-mediated pathway.

The survivin gene-silencing effects of different siRNA formulations were determined by real-time polymerase chain reaction (PCR) and enzyme-linked immunosorbent assay (ELISA) methods for mRNA and protein levels, respectively. As shown in FIGURE 4, mRNA and protein downregulation results are comparable. siRNA formulated in LPD-PEG and LPD-PEG-AA downregulated 30% and 70% survivin mRNA and protein, respectively, while

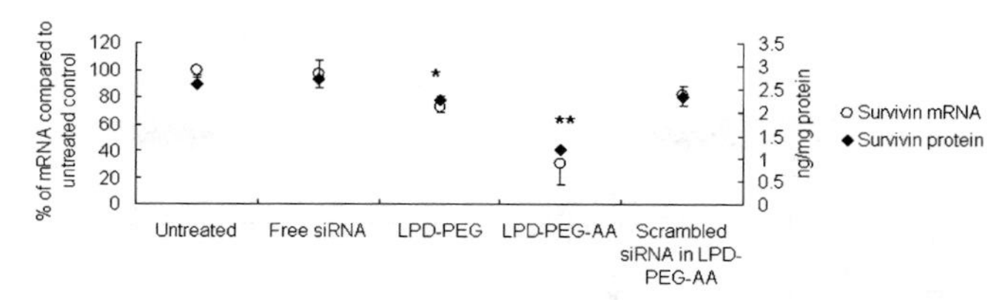

FIGURE 4. Survivin mRNA and protein levels in NCI-H1299 cells after siRNA treatment. NCI-1299 cells were treated with siRNA in different formulations at 37°C for 4 h. Survivin mRNA were quantified 24 h after treatment by real-time PCR and the data are expressed as a percentage of the untreated control. Survivin proteins were quantified 48 h after treatment with the ELISA method. Data = mean ± SD ($n = 3$). *indicates $P < 0.05$ and **indicates $P < 0.01$. (Reproduced with permission from Li and Huang, 2006.[7])

LPD-PEG-AA containing a scrambled siRNA showed a relatively low effect (20%, not significant). Free siRNA showed no effect.

Downregulation of survivin in NCI-1299 cells induced significant apoptosis determined by Annexin V staining assay (TABLE 1). Annexin V stained cells (apoptotic cells) were analyzed by flow cytometry. Annexin V positive cells treated with medium only, siRNA in PBS, siRNA in LPD-PEG, siRNA in LPD-PEG-AA, and scrambled siRNA in LPD-PEG-AA were 10%, 15%, 21%, 87%, and 38%, respectively.

In the chemosensitization study, we treated the cells with siRNA and then challenged the cells with different concentrations of cisplatin. Cell viability was determined by MTT assay. As shown in FIGURE 5, only siRNA in LPD-PEG-AA could sensitize NCI-1299 cells to cisplatin. The IC50 was reduced from 40 μM to 10 μM. The chemosensitization effect was highly sequence and formulation dependent, as the nontargeted LPD-PEG containing siRNA and the targeted LPD-PEG-AA containing a control RNA had no effect.

The results demonstrate that LPD-PEG-AA efficiently delivered siRNA into human lung cancer cells, downregulated survivin, and sensitized the cells to anticancer drugs.

TABLE 1. Flow cytometry analysis of FITC-Annexin V-stained NCI-1299 cells after treatment with siRNA

Treatment	Percentage of Annexin V positive*
Medium	10.64
siRNA in PBS	14.71
siRNA in LPD-PEG	20.72
siRNA in LPD-PEG-AA	87.43
Scrambled siRNA in LPD-PEG-AA	37.53

(Reproduced with permission from *Journal of Molecular Pharmaceutics* 2006.[7] Copyright 2006 Am. Chem.)

*Annexin V positive cells were analyzed based on 10,000 cells.

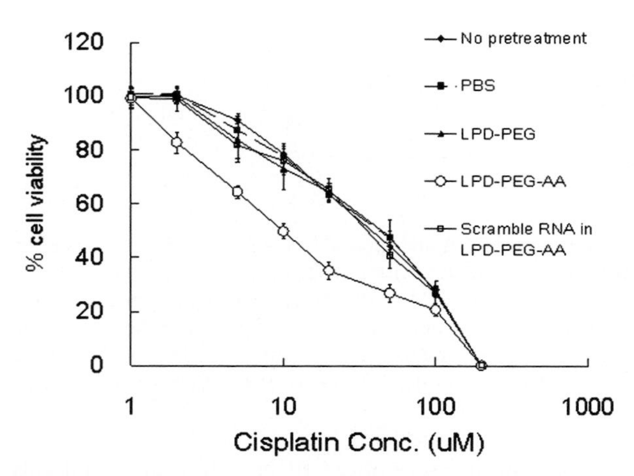

FIGURE 5. Chemosensitization of NCI-1299 cells mediated by pretreatment of siRNA in different formulations. Cells were incubated with siRNA in different formulations at 37°C for 4 h. Twenty-four hours later, cells were challenged with various concentrations of cisplatin. The viability of cells was measured 48 h later by MTT assay. Data = mean ± SD ($n = 3$). (Reproduced with permission from Li and Huang, 2006.[7])

TISSUE DISTRIBUTION STUDY

Human lung cancer cells (NCI-H460, 5×10^6) were subcutaneously injected into the right flank of athymic nude mice. Ten to 14 days later when the tumors reached the size of 1 cm × 1 cm, the mice were randomly grouped to receive intravenous injection of FAM-siRNA in PBS, LPD-PEG, or LPD-PEG-AA. Four hours after injection, the mice were sacrificed and major organs including hearts, livers, spleens, lungs, kidneys, and tumors were excised and imaged by the IVIS™ Imaging System (Xenogen Imaging Technologies, Alameda, CA).

FIGURE 6 shows representative images of the major organs. None of the organs except the liver from free FAM-siRNA-treated mice showed significant fluorescence. For mice treated with FAM-siRNA in LPD-PEG or LPD-PEG-AA, the tumor showed strong fluorescence and other normal organs showed only basal levels of signal. In this study, fluorescence signals indicate the presence of FAM-siRNA. Free siRNA showed very little tissue uptake, implying that free siRNA might be degraded and eliminated quickly from the blood before it is taken up by the organs. Surfaced-modified formulations were shown to deliver FAM-siRNA predominantly to the tumor, and there was no difference between nontargeted and targeted LPD formulations. This result suggests that tumor localization of the formulation is mainly dependent on the EPR effect of nanoparticles. Targeting ligands had no influence on the biodistribution of the nanoparticles. The reason for the surface-modified LPD delivering the majority of the dose to the tumor, instead of the liver, remains unclear. The possible

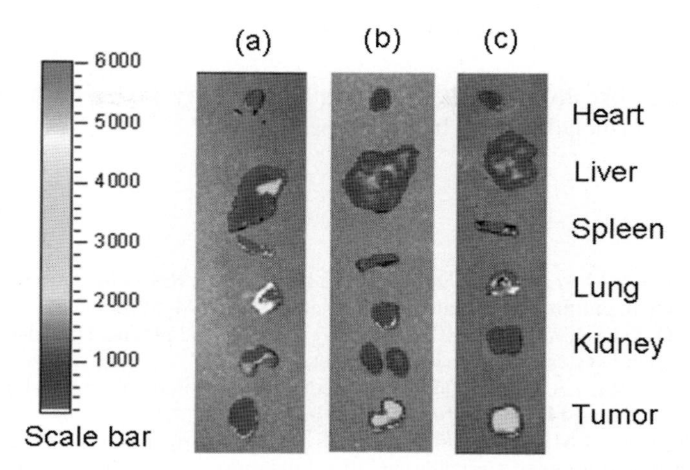

FIGURE 6. Fluorescence imaging on organs collected from NCI-H460 xenografted nude mice treated with free FAM-siRNA (**A**), FAM-siRNA in LPD-PEG (**B**), or FAM-siRAN in LPD-PEG-AA (**C**). (Shown in color in online version)

explanation may be the particle size of PEGylated LPD (120–150 nm) is too large to penetrate the liver sinusoidal fenestrae (pore size ∼ 100 nm)[8] but still small enough to extravasate through the capillaries in the tumor tissues (pore size ∼400 nm).[9] It was estimated that both LPD-PEG and LPD-PEG-AA could deliver about 70–80% of the injected dose per gram of organ weight.

PROSPECTS AND FUTURE PLAN

Systemic tumor-targeted delivery remains the most challenging issue in the drug delivery field. The ultimate goal is to deliver most of the injected dose into the tumor to achieve improved therapeutic effect and reduced side effect. Nanoparticle-based delivery systems serve this purpose due to the EPR effect featured by the highly vasculated tumors. However, only 1–15% of the injected dose accumulated in the tumors was accomplished so far by various delivery systems developed by different groups.[10–12] Nonspecific Reticular Endothelial System (RES) uptake contributes to the major loss of the administered dose.

Our ligand targeted, PEGylated LPD formulation showed significant increase in cellular uptake via specific receptor-mediated pathway. This targeted formulation also demonstrated its strong gene-silencing effect mediated by RNAi. Preliminary data showed that the surface-modified LPD delivered siRNA predominantly to the tumor, which was the major uptake organ, after intravenous administration. Our formulation provides an advantage of high tumor targeting and low RES uptake, which implied its potential for RNAi-based tumor therapy.

ACKNOWLEDGMENT

We would like to thank the Industrial Technology Research Institute in Taiwan for financially supporting this research.

REFERENCES

1. ELBASHIR, S.M. *et al.* 2001. Duplexes of 21-nucleotide RNAs mediate RNA interference in cultured mammalian cells. Nature **411:** 494–498.
2. LI, S.-D. & L. HUANG. 2006. Gene therapy progress and prospects: non-viral gene therapy by systemic delivery. Gene Ther. (in press).
3. ZIMMERMANN, T.S. *et al.* 2006. RNAi-mediated gene silencing in non-human primates. Nature **441:** 111–114.
4. SCHIFFELERS, R.M. *et al.* 2004. Cancer siRNA therapy by tumor selective delivery with ligand-targeted sterically stabilized nanoparticle. Nucleic Acids Res. **32:** e149.
5. BRANNON-PEPPAS, L. & J.O. BLANCHETTE. 2004. Nanoparticle and targeted systems for cancer therapy. Adv. Drug Deliv. Rev. **56:** 1649–1659.
6. YU, W. *et al.* 2004. A sterically stabilized immunolipoplex for systemic administration of a therapeutic gene. Gene Ther. **11:** 1434–1440.
7. LI, S.-D. & L. HUANG. 2006. Targeted delivery of antisense oligodeoxynucleotide and small interference RNA into lung cancer cells. Mol. Pharm. (in press).
8. BRAET, F. & E. WISSE. 2002. Structural and functional aspects of liver sinusoidal endothelial cell fenestrae: a review. Comp Hepatol. **1:** 1.
9. MONSKY, W.L. *et al.* 1999. Augmentation of transvascular transport of macromolecules and nanoparticles in tumors using vascular endothelial growth factor. Cancer Res. **59:** 4129–4135.
10. MAMOT, C. *et al.* 2005. Epidermal growth factor receptor-targeted immunoliposomes significantly enhance the efficacy of multiple anticancer drugs *in vivo.* Cancer Res. **65:** 11631–11638.
11. ROSSIN, R. *et al.* 2005. 64Cu-labeled folate-conjugated shell cross-linked nanoparticles for tumor imaging and radiotherapy: synthesis, radiolabeling, and biologic evaluation. J. Nucl. Med. **46:** 1210–1218.
12. DE WOLF, H.K. *et al.* 2005. In vivo tumor transfection mediated by polyplexes based on biodegradable poly(DMAEA)-phosphazene. J. Control Release **109:** 275–287.

Atelocollagen-Mediated Systemic DDS for Nucleic Acid Medicines

KOJI HANAI,[a] FUMITAKA TAKESHITA,[b] KIMI HONMA,[c]
SHUNJI NAGAHARA,[a] MIHO MAEDA,[a] YOSHIKO MINAKUCHI,[a]
AKIHIKO SANO,[a] AND TAKAHIRO OCHIYA[b]

[a]*Formulation Laboratories, Technology Research and Development Center, Dainippon Sumitomo Pharma Co., Ltd., Osaka 567-0878 Japan (Formerly; Formulation Research Laboratories, Sumitomo Pharmaceuticals Co., Ltd.)*

[b]*National Cancer Center Research Institute, Tokyo 104-0045 Japan*

[c]*Koken Bioscience Institute, Tokyo 115-0051 Japan*

ABSTRACT: The goal of our research is to provide a practical platform for drug delivery in oligonucleotide therapy. We report here the efficacy of an atelocollagen-mediated oligonucleotide delivery system applied to systemic siRNA and antisense oligonucleotide treatments in animal disease models. Atelocollagen and oligonucleotides formed a complex of nanosized particles, which was highly stable against nucleases. The complex allowed oligonucleotides to be delivered efficiently into several organs and tissues via intravenous administration. In a tumor metastasis model, the complex successfully delivered siRNA to metastasized tumors in bone tissue and inhibited their growth. We also demonstrated that a single intravenous treatment of the antisense oligodeoxynucleotide complex suppressed ear dermatitis in a contact hypersensitivity model. These results indicate the strong potential of the atelocollagen-mediated drug delivery system for practical therapeutic technology.

KEYWORDS: atelocollagen; drug delivery system; oligonucleotides; siRNA; antisense; systemic treatment

INTRODUCTION

Oligonucleotide therapies, using siRNAs and antisense oligodeoxynucleotides (ODNs), are anticipated as the next-generation therapeutic strategies for serious diseases. However, there are numerous challenges to be overcome, such as improving nuclease resistance, biodistribution, and cellular uptake of oligonucleotides, before the strategy is recognized as practical. The goal of our research is to provide a solution using a drug delivery system (DDS).

Address for correspondence: Koji Hanai, Formulation Laboratories, Technology Research and Development Center, Dainippon Sumitomo Pharma Co., Ltd., 3-45 Kurakakiuchi 1-chome, Ibaraki, Osaka 567-0878, Japan. Voice: +81-72-627-8146; fax: +81-72-627-8140.
e-mail: koji-hanai@ds-pharma.co.jp

Ann. N.Y. Acad. Sci. 1082: 9–17 (2006).
doi: 10.1196/annals.1348.010

Our DDS technology, atelocollagen-mediated DDS, has demonstrated effective delivery performances for many types of protein drugs.[1] Regarding the delivery of nucleic acid drugs, we have succeeded in applying DDS technology to almost any types of nucleic acid drugs, including siRNA,[2–4] antisense ODNs,[5–10] and plasmid DNA,[1,8,11] since our first evidence of enhancing the efficiency of gene delivery.[11] Apart from *in vitro* usage, atelocollagen-mediated DDS can be used for *in vivo* treatments as both local and systemic applications. Here, we describe recent research outcomes, particularly the systemic applications of siRNAs[2] and antisense ODNs[5] as well as the properties of the complex formed with atelocollagen and oligonucleotides.

COMPLEX FORMATION

Atelocollagen

Atelocollagen is a processed natural biomaterial produced from bovine type I collagen. It inherits useful biomaterial characteristics from collagen, such as a high biocompatibility and a high biodegradability.[12] The parts of collagen that are attributed to its immunogenicity, namely telopeptides, are eliminated in the process of atelocollagen production. Therefore, atelocollagen possesses little immunogenicity.[1] Indeed, it is widely used in various biodegradable medical, plastic surgical, and cosmetic supplies.

Complex Formation

When atelocollagen and an oligonucleotide, either siRNA or antisense ODN, are mixed with each other in a gentle manner, they form a complex. The size of the complex can be controlled between less than 200 nm and more than 10 μm, by adjusting the concentrations of the components, the type of additive used, and/or the handling procedure. Once the complex is formed, the oligonucleotides become stable against nucleases[3] (FIG. 1). The complex enhances oligonucleotide activity: for instance, FIGURE 2 demonstrates the inhibition of luciferase production by HEK 293 cells, to which the pGL3 plasmid has been introduced, by the transfection of a GL3 siRNA complex formed with atelocollagen. The inhibition ratio is as high as that of siRNA transfected with a cationic liposome reagent.[3]

SYSTEMIC APPLICATION OF ATELOCOLLAGEN-MEDIATED DDS TO siRNA DELIVERY

Profile of siRNA Delivery by Complex Formation

When a tumor (PC-3M-luc-6C cells) inoculated on the back of an athymic mouse reached a volume of 50–100 mm^3 (on day 8), an siRNA complex

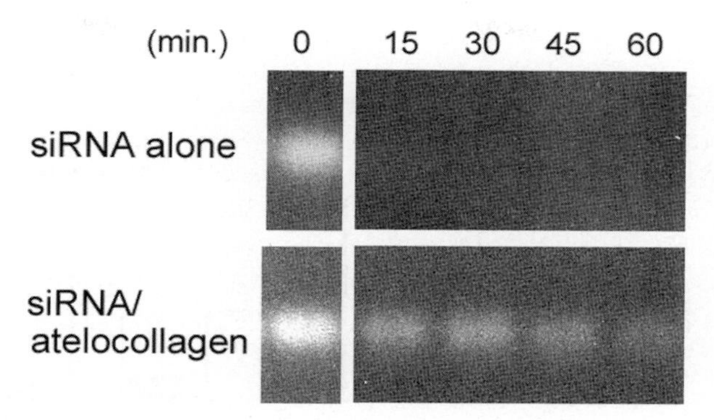

FIGURE 1. Stability of siRNA against nuclease by forming a complex with atelocollagen. siRNA alone (luciferase GL3 duplex) and siRNA complex formed with 0.5% atelocollagen were incubated in a solution containing 0.1 mg/mL RNase A at 37°C for up to 60 min. The extracted siRNA was agarose gel electrophoresed and visualized by ethidium bromide staining.

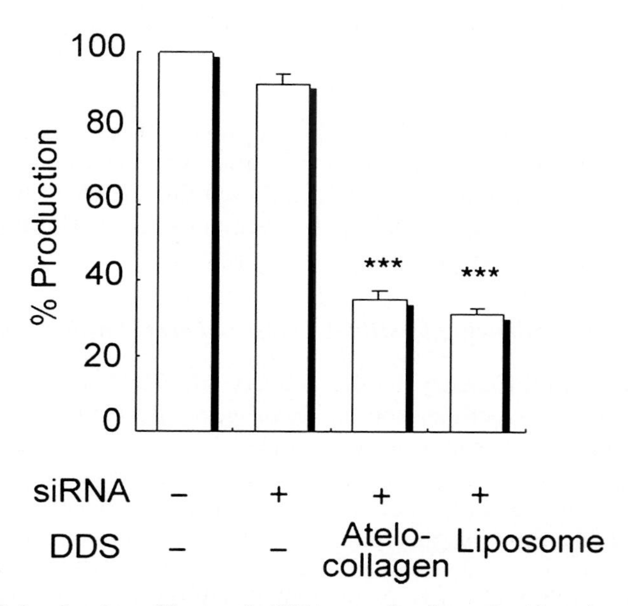

FIGURE 2. *In vitro* efficacy of siRNA complex formed with atelocollagen. GL3 siRNA alone or its complex with atelocollagen or cationic liposome was transfected into HEK 293 cells into which the pGL3 plasmid had been introduced. Luciferase activity was determined on day 2 ($n = 4$, mean ± S.E.). ***$P < 0.001$ versus siRNA-alone-treated cells.

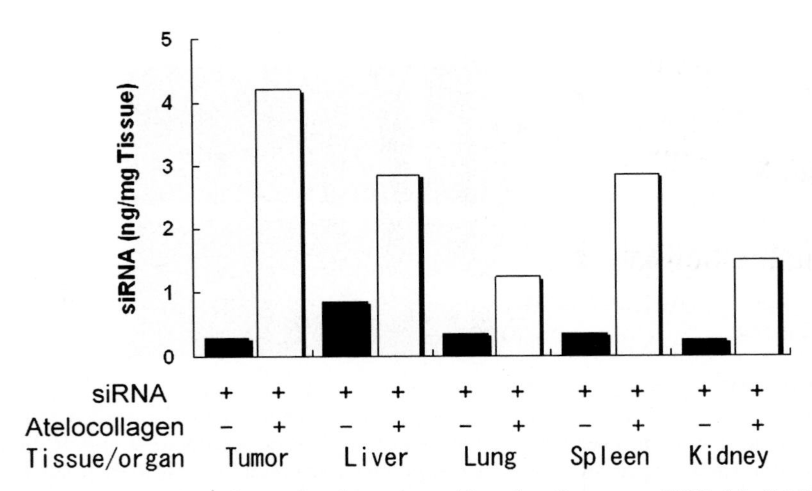

FIGURE 3. Effect of complex formation with atelocollagen on siRNA biodistribution. Two hundred microliters of luciferase GL3 siRNA alone (25 μg) or siRNA complex formed with atelocollagen (0.05%) was intravenously given to the tumor-bearing mice. One day after administration, total RNA was extracted from the tumor and selected organs. GL3 siRNA was detected by RNase protection assay. siRNA level was corrected for wet tissue weight.

formed with atelocollagen was injected intravenously. The tumor and selected organs were harvested and siRNA concentration was determined 24 h after the treatment. As a result, the complex allowed the siRNA to be delivered effectively into many organs, such as the liver and kidneys. However, siRNA was delivered most efficiently into the tumor (FIG. 3).

Luciferase Gene Silencing by siRNA Complex Formed with Atelocollagen

To evaluate the efficiency of the delivery of siRNA by atelocollagen-mediated DDS, an experiment on luciferase gene silencing was performed. In advance, animals were transplanted with PC-3M-luc-C6 tumor cells, which are bioluminescent human prostate carcinoma cells, by injection directly into bloodstream from the heart to develop a bone-metastatic cancer model.[13] Four weeks after transplantation, the mice were examined by bioluminescent imaging and metastasis was determined. Then, atelocollagen (0.05%) alone, luciferase GL3 siRNA (25 μg) alone, or their siRNA/atelocollagen complex was intravenously administered. The animals were reexamined 1 day after the treatment, and the pre- and posttreatment images were compared. Only the mice treated with the siRNA/atelocollagen complex showed an approximately 90% inhibition in bioluminescence (FIG. 4 A, B). The data indicate that the complex can efficiently deliver siRNA into metastatic tumor tissues and enhance *in vivo* siRNA activity.

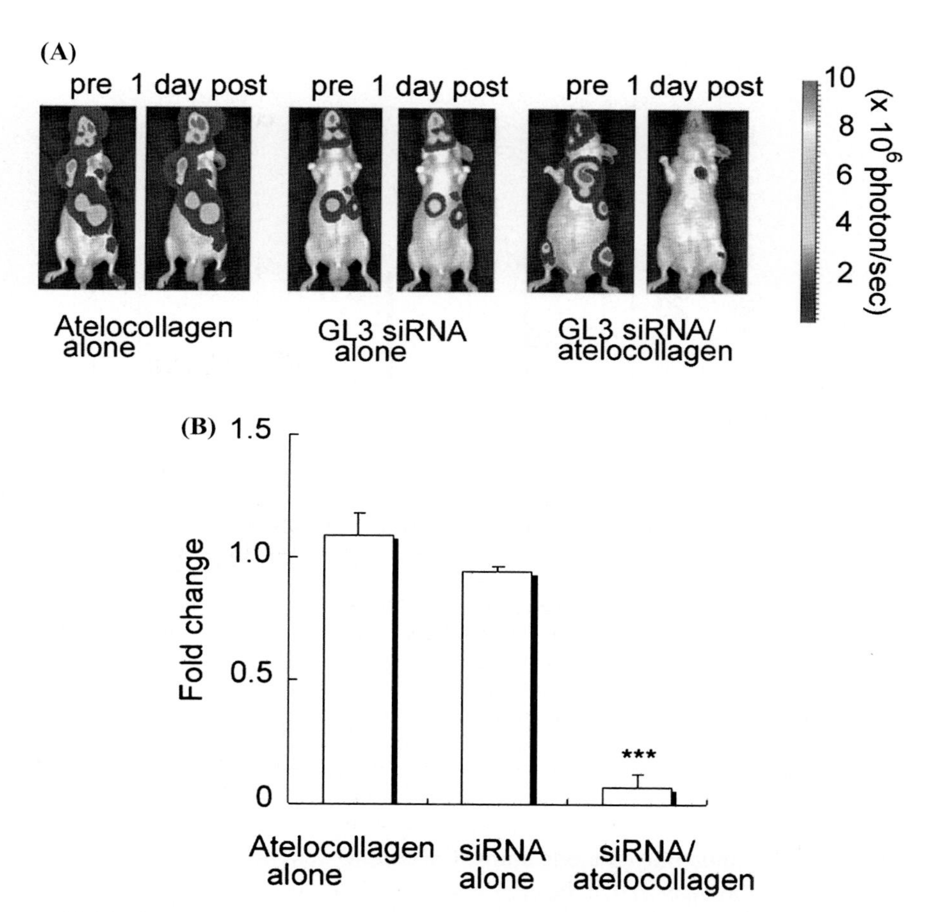

FIGURE 4. Inhibition of luciferase production by metastatic tumor. (**A**) Bioluminescent imaging of mice that suffered from tumor metastasis of PC-3M-luc-C6 cells. Two hundred microliters of atelocollagen alone (0.05%), siRNA alone (luciferase GL3 siRNA, 25 μg), or siRNA complex formed with atelocollagen was intravenously injected. Bioluminescent imaging was performed at pretreatment and 1 day after the treatment. (**B**) Group mean of bioluminescence strength ratio of posttreatment to pretreatment. Data represent the mean ± S.D. ($n = 4$). $^{***}P < 0.001$ versus siRNA-alone-treated group.

Inhibition of Tumor Growth at Metastatic Site

To evaluate the effect of the delivered siRNA on *in vivo* tumor growth inhibition, the following experiment was performed. As described above, metastatic cancers were developed. Repeated treatments of atelocollagen (0.05%) alone, the human enhancer of zeste homolog 2 (EZH2) siRNA (50 μg) alone, or their siRNA/atelocollagen complex were given intravenously 3, 6, and 9 days after the initiation of metastasis. The bioluminescence on day 28 in the groups

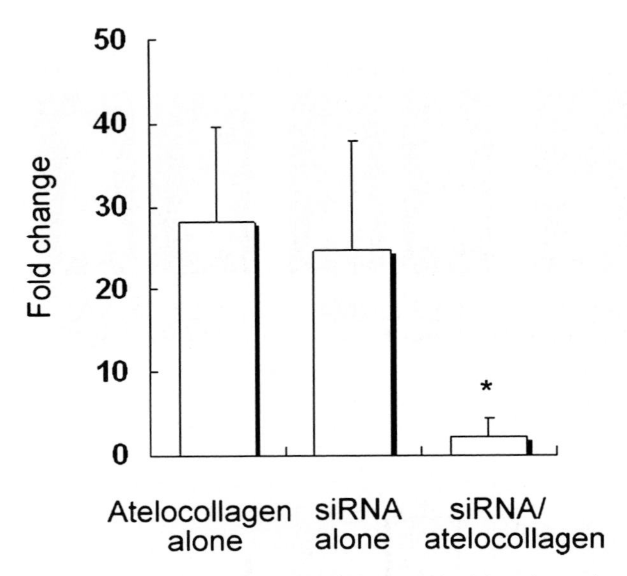

FIGURE 5. Suppression of tumor growth metastasized into bone tissue by siRNA complex formed with atelocollagen. PC-3M-luc-C6 tumor cells were injected into the bloodstream of mice to develop the tumor metastatic model. On days 3, 6, and 9, 200 μL of atelocollagen alone (0.05%), EZH2 siRNA alone (50 μg), or siRNA complex formed with atelocollagen was intravenously administered. The fold change was shown as normalized bioluminescence strength on day 28 (the value on day 28 divided by that on day 2 for each mouse). The data represent the mean ± S.D. ($n = 8$). $^*P < 0.05$ versus siRNA-alone-treated group.

treated with atelocollagen alone and siRNA alone was more than 25-fold that on the initial day (day 2). On the contrary, bioluminescence in the group treated with the siRNA complex formed with atelocollagen did not increase during the experiment (FIG. 5). The efficacy of the siRNA complex was also confirmed by histopathology (data not shown). Therefore, the systemic treatment using the siRNA complex formed with atelocollagen could be considered as a promising strategy for the inhibition of bone-metastatic prostate tumor growth *in vivo*.

SYSTEMIC APPLICATION OF ATELOCOLLAGEN-MEDIATED DDS TO ANTISENSE ODN DELIVERY

As an example of the systemic application of atelocollagen-mediated DDS to antisense ODN delivery, a study of suppressing inflammatory reactions was conducted. An antimurine-ICAM-1 antisense ODN (ISIS 3082) was used for the study.[14] The antisense ODN delivery performance was assessed using a well-known contact hypersensitivity model in mice.[14] Briefly, the animals were sensitized with 0.5% 2,4-dinitrofluorobenzene (DNFB) for 2 consecutive

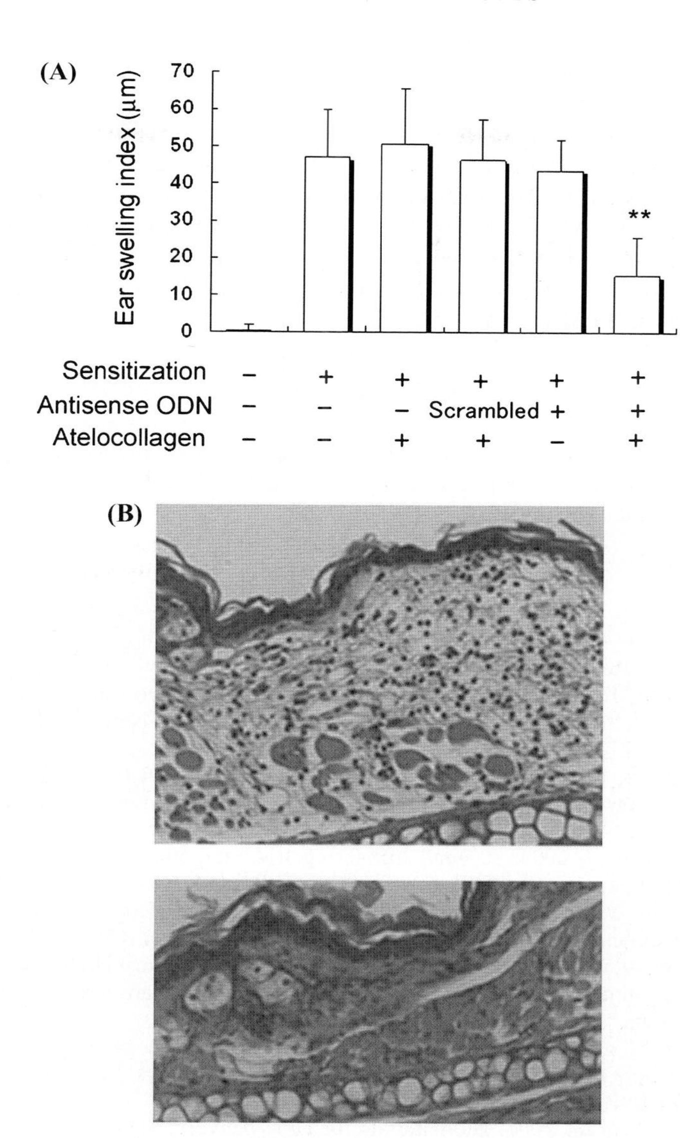

FIGURE 6. Suppression of inflammatory reaction by antisense ODN complex formed with atelocollagen. Ear dermatitis was provoked in accordance with a previous study.[14] Antisense ODN (0.6 mg/kg) formulated with atelocollagen (0.05%) was intravenously administered 15 min after the ear challenge. Ear swelling was measured and the ear pinna was sampled for histopathology 24 h later. (**A**) The ear swelling index, which was calculated by subtracting the prechallenge ear thickness from the postchallenge ear thickness. Data represent the mean $\pm$ SD ($n = 6$). $^{**}P < 0.01$ versus sensitized control group. (**B**) Histopathologic images. The top and bottom figures show the ears of a sensitized control and an animal treated with antisense ODN complex formed with atelocollagen, respectively. Hematoxylin and eosin staining. The pictures are presented at the same magnification.

days, and given a dab of 0.2% DNFB on the ear surface as an ear challenge on day 7. A single intravenous injection of antisense ODN (0.6 mg/kg) complex formed with atelocollagen (0.05%) 15 min after the ear challenge successfully reduced the ear swelling at 24 h (FIG. 6 A), while antisense ODN alone failed to suppress the swelling. These results were supported by histopathologic examinations that showed lesser dermatitis defined by slight edema and inflammatory cell infiltration in the antisense ODN complex group (FIG. 6 B). Moreover, immunohistochemistry demonstrated that the complex downregulated ICAM-1 expression (data not shown).

CONCLUSION

The results described here demonstrate that atelocollagen-mediated DDS has a great potential for practical application to systemic treatments using oligonucleotides *in vivo*.

REFERENCES

1. SANO, A. *et al.* 2003. Atelocollagen for protein and gene delivery. Adv. Drug Delivery Rev. **55:** 1651–1677.
2. TAKESHITA, F. *et al.* 2005. Efficient delivery of small interfering RNA to bone-metastatic tumors by using atelocollagen in vivo. Proc. Natl. Acad. Sci. USA **102:** 12177–12182.
3. MINAKUCHI, Y. *et al.* 2004. Atelocollagen-mediated synthetic small interfering RNA delivery for effective gene silencing in vitro and in vivo. Nucleic Acid Res. **32:** e109.
4. TAKEI, Y. *et al.* 2004. A small interfering RNA targeting vascular endothelial growth factor as cancer therapeutics. Cancer Res. **64:** 3365–3370.
5. HANAI, K. *et al.* 2004. Potential of atelocollagen-mediated systemic antisense therapeutics for inflammatory disease. Human Gene Ther. **15:** 263–272.
6. HIRAI, K. *et al.* 2003. Antisense oligodeoxynucleotide against HST-1/FGF-4 suppresses tumorigenicity of an orthotopic model for human germ cell tumor in nude mice. J. Gene Med. **5:** 951–957.
7. HONMA, K. *et al.* 2001. Atelocollagen-based gene transfer in cells allows high-throughput screening of gene functions. Biochem. Biophys. Res. Commun. **289:** 1075–1081.
8. OCHIYA, T. *et al.* 2001. Biomaterials for gene delivery: atelocollagen-mediated controlled release of molecular medicines. Curr. Gene Ther. **1:** 31–52.
9. TAKEI, Y. *et al.* 2001. Antisense oligodeoxynucleotide targeted to Midkine, a heparin-binding growth factor, suppresses tumorigenicity of mouse rectal carcinoma cells. Cancer Res. **61:** 8486–8491.
10. TAKEI, Y. *et al.* 2002. 5'-, 3'-Inverted thymidine-modified antisense oligodeoxynucleotide targeting midkine. J. Biol. Chem. **277:** 23800–23806.
11. OCHIYA, T. *et al.* 1999. New delivery system for plasmid DNA in vivo using atelocollagen as a carrier material: the minipellet. Nat. Med. **5:** 707–710.
12. MIYATA, T. *et al.* 1992. Collagen engineering for biomaterial use. Clin. Mater. **9:** 139–148.

13. JENKINS, D.E. *et al.* 2003. In vivo monitoring of tumor relapse and metastasis using bioluminescent PC-3M-luc-C6 cells in murine models of human prostate cancer. Clin. Exp. Metastasis **20:** 745–756.
14. KLIMUK, S.K. *et al.* 2000. Enhanced anti-inflammatory activity of a liposomal intracellular adhesion molecule-1 antisense oligodeoxynucleotide in an acute model of contact hypersensitivity. J. Pharmacol. Exp. Ther. **292:** 480–488.

Intracellular Delivery of Oligonucleotide Conjugates and Dendrimer Complexes

R.L. JULIANO

Department of Pharmacology, School of Medicine, University of North Carolina, Chapel Hill, North Carolina 27599, USA

ABSTRACT: Enhancing the delivery of antisense and siRNA molecules to cells and tissues is a key issue for oligonucleotide therapeutics. Cell-penetrating peptides (CPPs) have the ability to convey linked "cargo" molecules into the cytosol; thus we have explored the use of CPPs as delivery agents for oligonucleotides. We have extensively evaluated CPP–oligonucleotide conjugates, and have recently begun to explore the use of CPP–dendrimer–oligonucleotide complexes. We have found that CPP-antisense oligonucleotide conjugates can be taken up by cells and can effectively modify gene expression in cell culture and in tissues. Although not as potent in cell culture as cationic lipid delivery agents, CPP–oligonucleotide conjugates offer the advantage of being molecules rather than particles, and may have substantial advantages over particle-based delivery in the *in vivo* setting.

KEYWORDS: antisense, siRNA, oligonucleotide, cell-penetrating peptides, CPP–dendrimer–oligonucleotide

INTRODUCTION

The advent of RNA interference as well as the emergence of new chemistries for base, sugar, and backbone modifications of antisense molecules has reenergized the field of oligonucleotide therapeutics.[1,2] Most of the experience to date at the level of therapy in animal models and in clinical trials resides with antisense oligonucleotides,[3] however, *in vivo* studies with siRNAs are also progressing rapidly.[4] The pharmacokinetics and biodistribution behavior of oligonucleotide drugs in the body can be considered in terms of a series of barriers to successful delivery of the drug to its intracellular locus of action. Therefore, even after entry into the blood compartment, the oligonucleotide must still surmount obstacles including (*a*) rapid renal excretion, (*b*) degradation by nucleases in blood and tissues, (*c*) unproductive uptake by phagocytes in liver, spleen, and elsewhere, (*d*) passage across the vascular endothelium,

Address for correspondence: R.L. Juliano, Department of Pharmacology, School of Medicine, University of North Carolina, Chapel Hill, NC 27599. Voice: 919-966-4383; fax: 919-966-5640.
e-mail: arjay@med.unc.edu

Ann. N.Y. Acad. Sci. 1082: 18–26 (2006).
doi: 10.1196/annals.1348.011

(*e*) diffusion through the extracellular matrix, and finally (*f*) passage across the plasma membrane or endosomal membranes so as to gain access to the cytoplasm.

There is an interesting dichotomy in studies of oligonucleotide delivery. Thus, in cell culture models of antisense and siRNA effects, it is essential that a delivery agent such as a cationic lipid be complexed with the oligonucleotide; there is minimal entry of "free" (uncomplexed) oligonucleotide into the cytoplasm or nucleus and thus minimal pharmacological effect. In contrast, there are numerous examples of *in vivo* studies where pharmacological effects have been attained with "free" antisense or siRNA oligonucleotides.[3,4] The reasons for this are unclear. One possibility is that a key transport system for oligonucleotides that exists in tissues is uniformly lost in cells in culture. Another possibility is that a different pharmacokinetic regime prevails *in vivo*, where the use of multiple doses of oligonucleotide over an extended period of time allows gradual accumulation at key intracellular sites.[5] In any case, the fact that most *in vivo* studies to date have utilized "free" oligonucleotides does not preclude the possibility that appropriate delivery approaches might further enhance the therapeutic benefits of antisense or siRNA oligonucleotides. It is this concept that has motivated our work on oligonucleotide delivery.

A variety of potential oligonucleotide delivery approaches are available. One of the most common strategies for nucleic acid delivery in general is to complex the nucleic acid with cationic lipids. In the case of plasmid DNA the resulting complexes (lipoplexes) have been extensively studied for *in vivo* applications in the gene therapy field.[6] Another approach has been to incorporate either plasmid DNA or oligonucleotides into polymeric nanoparticles of various types.[7,8] However, a major disadvantage of both these approaches for oligonucleotide therapeutics is that the oligonucleotide, initially a mid-sized molecule with a potentially broad tissue distribution, is converted into a particulate entity with a much more limited ability to access many tissues. Thus, in thinking about oligonucleotide delivery strategies we have focused on two approaches that involve delivery moieties that are molecular rather than particulate in nature. The two approaches we have pursued are: (*a*) conjugates between oligonucleotides and "cell-penetrating peptides" and (*b*) complexes between oligonucleotides and dendrimers.

RESULTS AND DISCUSSION

The term "cell-penetrating peptide" (CPP) covers several families of peptides that have been identified over the last few years. While some CPPs comprise largely of hydrophobic residues, many are polycationic in nature with multiple arginines seemingly important to their function.[9,10] CPPs have the ability to readily enter cells by passage across the plasma membrane or endosomal membranes, and in doing so they can often carry a "cargo" molecule into the cytosol as well. CPPs have been used very effectively for the delivery of peptides, some oligonucleotides, and some proteins, although delivery of large, complex molecules is not always assured.[11] Two of the prototypical CPPs are those derived from the HIV TAT and *drosophila* Antennepedia transcriptional regulators. We have made extensive use of oligonucleotides chemically conjugated to the "TAT" and "ANT" peptide sequences and have evaluated their intracellular delivery and pharmacological effects in cells in culture and, to a limited extent, in animals.

In our studies we have made use of two types of model systems to test oligonucleotide delivery. Our first model involves the *MDR1* gene as a target. *MDR1* codes for the P-glycoprotein, a membrane transporter that pumps a variety of drug molecules out of cancer cells and thus confers a multidrug-resistant phenotype. In highly drug-resistant cells *MDR1* is amplified, message levels are high, and P-glyoprotein is abundant and turns over slowly.[12] Thus this model is a very challenging one for oligonucleotide-mediated inhibition of gene expression. Our second model, developed by our UNC colleague R. Kole, involves use of splice-switching oligonucleotides.[13] Thus an abnormal intron is placed within the coding sequence of a reporter such as luciferase or GFP resulting in abnormal splicing and a message that does not given rise to a functional protein. The abnormal splicing can be overcome by the presence of an antisense oligonucleotide that targets the mutant splice junction, thus correcting the process and resulting in the production of the reporter gene protein. An advantage of this approach is that it provides a positive "read-out" for antisense delivery and action.

We have used CPP–oligonucleotide conjugates with both of the models described above.[14–16] Typically, the peptide–oligonucleotide conjugates are synthesized including an S–S linkage that should be bioreversible in the SH-rich environment of the cytosol; the conjugates also include a 3′-fluorophore to permit easy visualization in cells (FIG. 1). After synthesis, the conjugates are purified by HPLC and then used in various types of experiments. Depending on their specific composition, peptide–oligonucleotide conjugates typically have masses of about 6–10 kDa and are thus of moderate molecular size. Peptide–oligonucleotide synthesis was done in collaboration with the laboratory of Dr. B. Shaw at Duke University. Although we have worked primarily with antisense oligonucleotides thus far, relatively minor adaptations of this approach could be used with siRNAs.

We have compared the intracellular delivery of CPP–oligonucleotide conjugates to that for the same oligonucleotide complexed with cationic lipids. As seen in FIGURE 2, there was substantial cell uptake of the conjugate, with fluorescence observed in the nucleus and in cytoplasmic vesicles. The lipid complex provided a greater degree of total cell uptake, while a "free" oligonucleotide did not visibly accumulate in cells. These results are typical of experiments with several types of cells and conjugates. Cellular uptake of the conjugates resulted in appropriate pharmacological effects. In the study

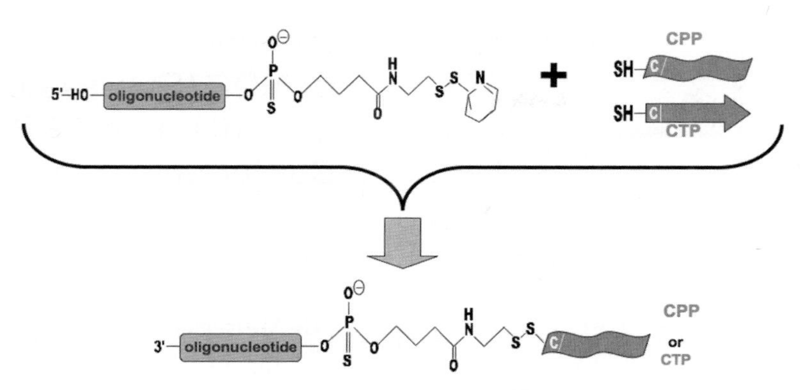

FIGURE 1. Preparation of peptide–oligonucleotide conjugates. Scheme for conjugation of a cell-penetrating peptide or a cell-targeting peptide to an oligonucleotide via a bioreversible S–S linkage. (Shown in color in online version)

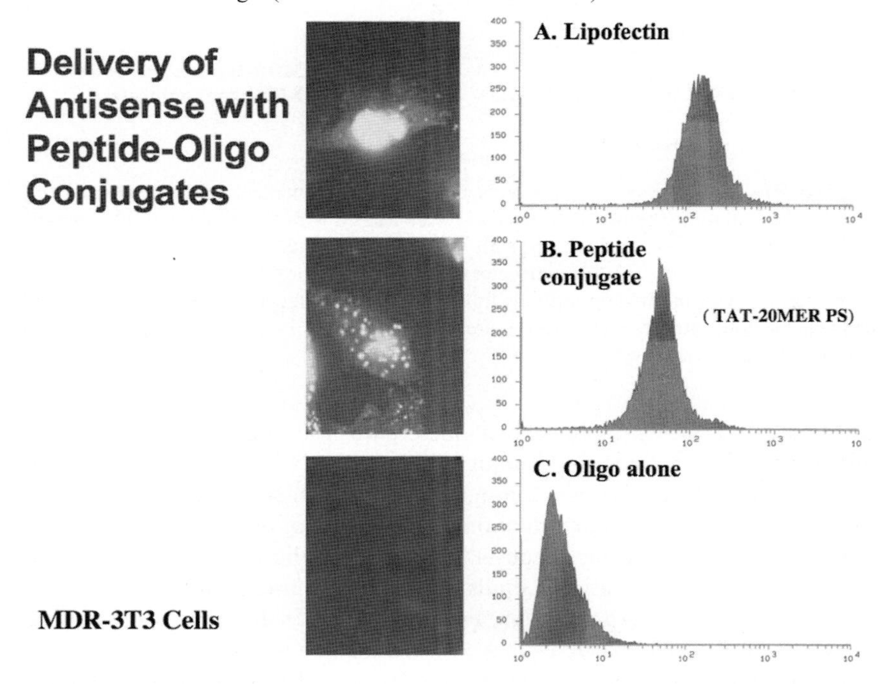

FIGURE 2. Cell uptake of peptide–oligonucleotide conjugates. An anti-MDR1 oligonucleotide was conjugated to the TAT peptide with further 3′ conjugation to a fluorophore. Cellular uptake of the conjugate (**B**) was compared to uptake of the same fluor-tagged oligonucleotide complexed with lipofectin (**A**) or used in free form (**C**). The left panels show images obtained by fluorescence microscopy whereas the right panels show data from flow cytometry (X-axis, fluorescence intensity; Y-axis, cell number). (Shown in color in online version)

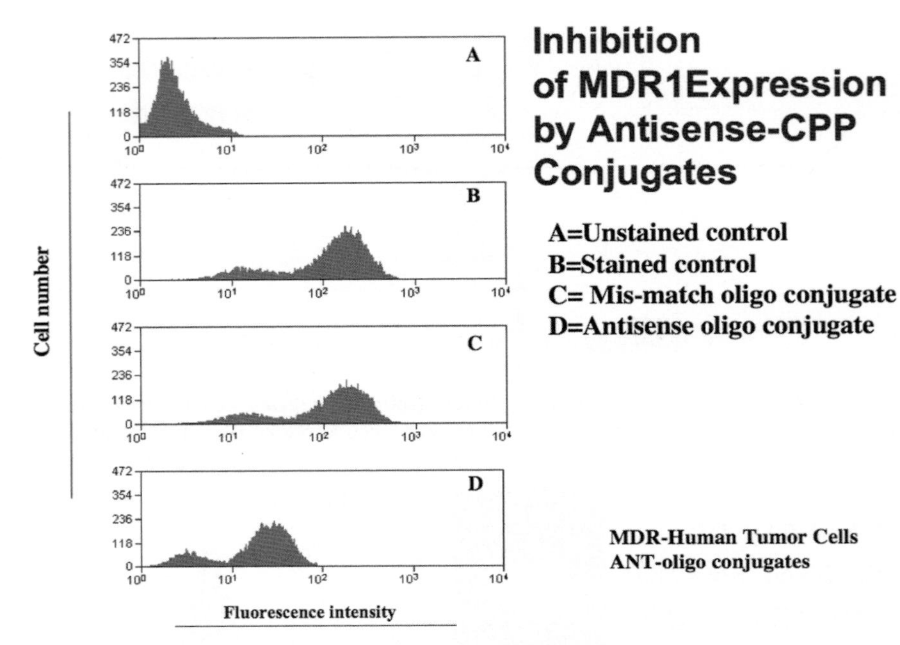

FIGURE 3. Pharmacological effects of peptide–oligonucleotide conjugates. Drug-resistant human tumor cells (MES-SA/Dx5) were treated with a conjugate of ANT peptide and anti-MDR1 oligonucleotide, or with an ANT–mismatch oligonucleotide conjugate. Cell surface expression of P-glycoprotein, the MDR1 gene product, was monitored by immunostaining with an anti-Pgp antibody followed by a fluor-tagged secondary antibody, with analysis by flow cytometry (*X*-axis, fluorescence intensity; *Y*-axis, cell number). (Shown in color in online version).

illustrated in (FIG. 3), a phosphorothioate antisense oligonucleotide directed against MDR1 message and conjugated to the ANT peptide was used. As indicated the ANT–antisense conjugate, but not a mismatched control conjugate, provided a substantial reduction in expression of the P-glycoprotein. Our overall experience thus far suggests that the CPP–oligonucleotide conjugates are less effective than cationic lipids in affording intracellular delivery in the cell culture context. Against this, however, the fact that the conjugates are molecules rather than particles may afford advantages in the *in vivo* context.

In this light we wished to examine the effects of CPP-oligonucleotide conjugates *in vivo*. These studies used an all methoxyethyl oligonucleotide provided by ISIS Pharmaceuticals (Carlsbad, CA); this was conjugated to the TAT cell-penetrating peptide. The tests were conducted in an animal model developed by R. Kole.[17] The conjugate was designed to correct splicing in a GFP gene having an inserted abnormal intron; this gene was ubiquitously present in tissues of a specially constructed transgenic mouse line. Thus, splice-switching by the oligonucleotide could be detected by either observing the accumulation

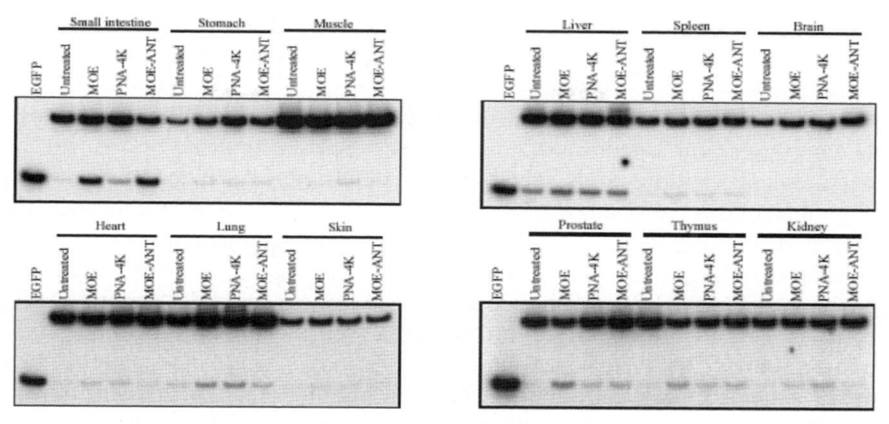

FIGURE 4. *In vivo* effects of splice-correcting peptide–oligonucleotide conjugates. This study used transgenic mice harboring an EGFP gene with an inserted abnormal intron that prevents correct splicing and thus inhibits EGFP expression. Splicing is restored when an antisense oligonucleotide binds to the abnormal splice junction. The mice were treated intraperitoneally with a methoxyethyl oligonucleotide, its ANT-conjugate, or a PNA–oligonucleotide with four lysines conjugated, all at 50 mg/kg. Tissues were harvested after 3 days and analyzed for correctly spliced EGFP message by quantitative PCR. (Shown in color in online version)

of GFP in tissues, or by measuring the amount of correctly spliced message by quantitative polymerase chain reaction (PCR). As seen in (FIG. 4), the TAT–oligonucleotide conjugate was moderately effective in splice correction in some tissues, as measured by PCR. However, in this experiment it was no more effective than the unconjugated methoxyethyl oligonucleotide. While this was slightly disappointing, we still have a lot to learn about the pharmacokinetics and biodistribution of CPP–oligonucleotide conjugates, and this information may lead to more efficacious use of these molecules. At a minimum, these experiments clearly demonstrated that peptide–oligonucleotide conjugates can have interesting biological effects in an animal model.

In addition to effectiveness, an important issue for oligonucleotide conjugates (or indeed any form of oligonucleotide) is the degree of selectivity; does the oligonucleotide primarily affect its intended target or are there many "off-target" actions? We approached this issue by using DNA array analysis.[14] Thus, cells treated with the CPP–oligonucleotide conjugates used in (FIG. 3), or with oligonucleotide alone or CPP alone, were compared using a gene array. In addition to inhibiting expression of MDR1, the target gene, approximately 2% of the genes in the array were affected. In some cases expression was reduced, whereas in others it was increased; the genes affected did not necessarily have any sequence identity with the oligonucleotide used. Effects on gene expression were observed with both the conjugates and the parent-"free" oligonucleotides (delivered by scrape-loading). These results suggest that antisense oligonucleotides or their peptide conjugates are reasonably selective (only 2% of genes affected), but are clearly not completely specific in their actions.

In addition to CPPs, we have also explored the delivery of oligonucleotides as complexes with dendrimers, these being a class of branched polymers.[18] In contrast to cationic lipids, when cationic dendrimers form complexes with oligonucleotides the size range of the complexes is in the tens to hundreds of thousands daltons range[19]; in other words the complexes are macromolecules and not particles. Our early experience with cationic dendrimers of the polyamidoamine (PAMAM) variety indicated that dendrimers could deliver antisense oligonucleotides to cells in culture about as well as common cationic lipid transfection agents.[20] However, in general we find PAMAM dendrimers to be rather toxic. This has led to the concept of using oligonucleotide dendrimers. In this scenario, an antisense oligonucleotide with a specific target would be stabilized by the inclusion of chemical modifications and then hybridized to a branched oligonucleotide ("oligo dendrimer") having complementary sequences comprising relatively unstable phosphodiester residues. Upon entry into the cells, the unstable branched oligomer would be degraded leading to release of the active antisense molecule.[16] To attain delivery, the branched oligomer would be conjugated to one or more CPPs. We have tested this concept in a preliminary manner. Thus, oligonucleotide dendrimers complexed to

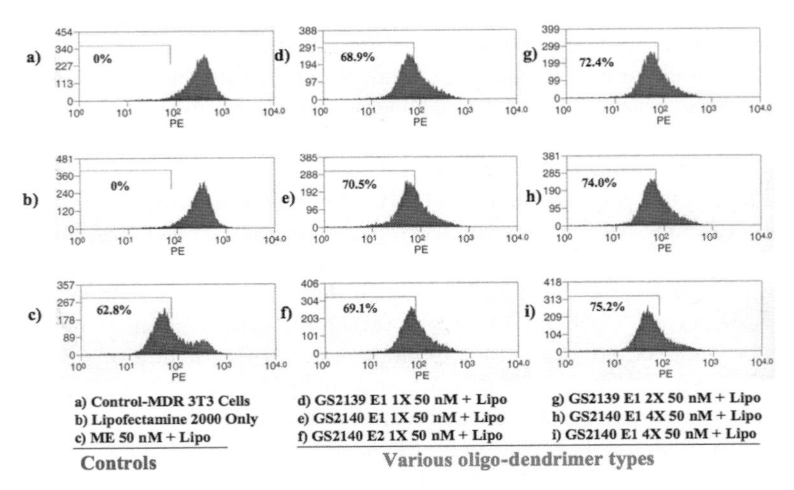

FIGURE 5. Inhibition of cell surface expression of P-glycoprotein using oligonucleotide dendrimers. Various oligonucleotide dendrimers[21] were complexed to a methoxyethyl gapmer anti-MDR1 oligonucleotide and then to Lipofectamine 2000. Multidrug-resistant 3T3 cells were treated with the complexes and the cell surface expression of P-glycoprotein was measured by immunostaining and flow cytometery. The percentages on the charts indicate the percentage of cells shifted to one standard deviation below the mean of the untreated controls.

an anti-MDR1 oligonucleotide were very effective in reducing P-glycoprotein expression when the complexes were delivered to cells with cationic lipids (FIG. 5).[21] The next step is to build in the CPP delivery moiety so as to avoid use of cationic lipids; this is in progress.

In summary, conjugates and complexes based on cell-penetrating peptides have shown some promise as delivery agents for oligonucleotides, both in cell culture and *in vivo*. The use of CPP–oligonucleotide conjugates seems particularly interesting, as these molecules are of moderate molecular size and may be able to attain broad tissue distribution. However, not all studies with peptide–oligonucleotide conjugates have been successful, and much remains to be learned about their biodistribution, pharmacology, and toxicology.[22]

REFERENCES

1. DORSETT, Y. & T. TUSCHL. 2004. siRNAs: applications in functional genomics and potential as therapeutics. Nat. Rev. Drug Discov. **3:** 318–329.
2. KURRECK, J. 2003. Antisense technologies. Improvement through novel chemical modifications. Eur. J. Biochem. **270:** 1628–1644.
3. CROOKE, S. 2004. Progress in antisense technology. Annu. Rev. Med. **55:** 61–95.
4. SOUTSCHEK, *et al.* 2004. Therapeutic silencing of an endogenous gene by systemic administration of modified siRNAs. Nature **432:** 173–178.
5. JULIANO, R. *et al.* 1999. Antisense pharmacodynamics: critical issues in the transport and delivery of antisense oligonucleotides. Pharm. Res. **16:** 494–502.
6. LIU, F. & L. HUANG. 2002. Development of non-viral vectors for systemic gene delivery. J. Control Release **78:** 259–266.
7. PANYAM, J. & V. LABHASETWAR. 2003. Biodegradable nanoparticles for drug and gene delivery to cells and tissue. Adv. Drug Deliv. Rev. **55:** 329–347.
8. VINOGRADOV, S., E. BATRAKOVA & A. KABANOV. 2004. Nanogels for oligonucleotide delivery to the brain. Bioconjug. Chem. **15:** 50–60.
9. WADIA, J. & S. DOWDY. 2002. Protein transduction technology. Curr. Opin. Biotechnol. **13:** 52–56.
10. JARVER, P. & U. LANGEL. 2004. The use of cell-penetrating peptides as a tool for gene regulation. Drug Discov. Today **9:** 395–402.
11. DIETZ, G. & M. BAHR. 2004. Delivery of bioactive molecules into the cell: the Trojan horse approach. Mol. Cell Neurosci. **27:** 85–131.
12. GOTTESMAN, M., T. FOJO & S. BATES. 2002. Multidrug resistance in cancer: role of ATP-dependent transporters. Nat. Rev. Cancer **2:** 48–58.
13. SAZANI, P. & R. KOLE. 2003. Therapeutic potential of antisense oligonucleotides as modulators of alternative splicing. J. Clin. Invest. **112:** 481–486.
14. FISHER, A. *et al.* 2002. Evaluating the specificity of antisense oligonucleotide conjugates: a DNA array analysis. J. Biol. Chem. **277:** 22980–22984.
15. ASTRIAB-FISHER, A. *et al.* 2002. Conjugates of antisense oligonucleotides with the Tat and antennapedia cell-penetrating peptides: effects on cellular uptake, binding to target sequences, and biologic actions. Pharm. Res. **19:** 744–754.
16. ASTRIAB-FISHER, A. *et al.* 2004. Increased uptake of antisense oligonucleotides by delivery as double stranded complexes. Biochem. Pharmacol. **68:** 403–407.

17. SAZANI, P. *et al.* 2002. Systemically delivered antisense oligomers upregulate gene expression in mouse tissues. Nat. Biotechnol. **20:** 1228–1233.
18. ESFAND, R. & D. TOMALIA. 2001. Poly(amidoamine) (PAMAM) dendrimers: from biomimicry to drug delivery and biomedical applications. Drug Discov. Today **6:** 427–436.
19. DELONG, R. *et al.* 1997. Characterization of complexes of oligonucleotides with polyamidoamine starburst dendrimers and effects on intracellular delivery. J. Pharm. Sci. **86:** 762–764.
20. YOO, H. & R. JULIANO. 2000. Enhanced delivery of antisense oligonucleotides with fluorophore-conjugated PAMAM dendrimers. Nucleic Acids Res. **28:** 4225–4231.
21. CHALTIN, P. *et al.* 2005. Delivery of antisense oligonucleotides using cholesterol-modified sense dendrimers and cationic lipids. Bioconjug. Chem. **16:** 827–836.
22. JULIANO, R. 2005. Peptide-oligonucleotide conjugates for the delivery of antisense and siRNA. Curr. Opin. Mol. Ther. **7:** 132–136.

Mechanism of PNA Transport to the Nuclear Compartment

LENKA STANKOVA, AMY J. ZIEMBA, ZHANNA V. ZHILINA, AND SCOT W. EBBINGHAUS

Arizona Cancer Center, University of Arizona, Tucson, Arizona 85724–5024, USA

ABSTRACT: We evaluated the nuclear uptake of fluorescently labeled peptide nucleic acids and measured the binding of unlabeled peptide nucleic acids (PNAs) to the endogenous HER-2/neu promotor in digitonin-permeabilized SK-BR-3 cells. Fluorescently labeled PNAs readily enter the nucleus of digitonin-permeabilized cells, and binding to the chromosomal target sequence was detected with a bis-PNA. Nuclear uptake and target sequence binding were inhibited by N-ethylmaleimide (NEM) and GTPγS. We conclude that PNAs are transported into the nucleus through an energy-dependent process involving the nuclear pore complex.

KEYWORDS: nuclear pore complex; oligonucleotide; peptide nucleic acid; N-ethylmaleimide; NEM; GTPγS; digitonin; permeabilized cells; SK-BR-3 cells; karyopherins

Peptide nucleic acids (PNAs) are DNA mimics that bind to DNA and RNA high affinity and resist degradation by nucleases. PNAs are inefficiently transported into cells, and carrier molecules are needed to mediate cellular uptake.[1] The subcellular localization of a PNA conjugated to a cell penetrating peptide (CPP) with and without a nuclear localization signal (NLS) showed that PNAs without an NLS did not enter the nuclear compartment.[2] PNAs that target DNA or pre-mRNA require efficient uptake in the nucleus, and we sought to evaluate the mechanism of PNA import into the nucleus using antigene mono- and bis-PNAs designed to target the HER-2/neu promotor.[3]

Experimentally, active transport through the nuclear pore complex (NPC) is inhibited by compounds such as N-ethylmaleimide (NEM), which binds to a cytosolic karyopherin, and GTPγS, which inhibits the Ran-GTPase.[4] Cells permeabilized with digitonin can be used to study the nuclear transport of molecules from the cytosol, and cytoplasmic extracts can enhance the uptake of some molecules.[5,6] In these studies, we studied the uptake and subcellular localization of fluorescently labeled PNAs in unfixed, adherent,

Address for correspondence: Scot W. Ebbinghaus, 1515 North Campbell Avenue, Tucson, AZ 85724–5024. Voice: 520-626-3424; fax: 520-626-5462.
e-mail: sebbinghaus@azcc.arizona.edu

Ann. N.Y. Acad. Sci. 1082: 27–30 (2006). © 2006 New York Academy of Sciences.
doi: 10.1196/annals.1348.064

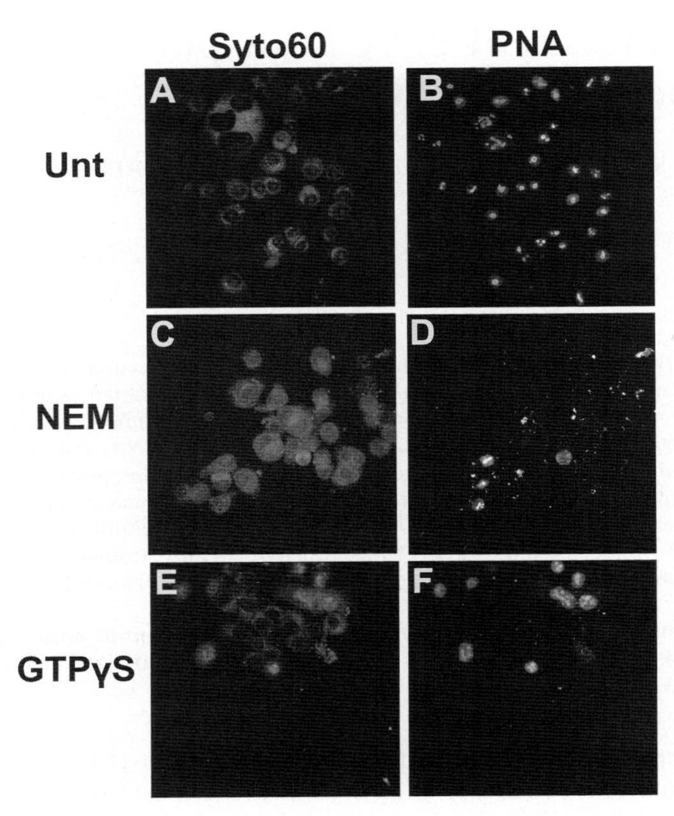

FIGURE 1. SK-BR-3 breast cancer cells were permeabilized using digitonin (30 μg/mL) and treated with 1μM PNA (TAMRA-CTCCTCCTC-O-KK) for 30 min, and the cells were counterstained using SYTO60 (**A, C, E**). Mid-section images of the cells showing nuclear uptake of the PNA under 40X magnification were generated using Z-series analysis (**B, D, F**). Representative images were taken from untreated cells (**A, B**), cells pretreated with NEM (**C, D**), and cells treated with GTPγS (**E, F**). Nuclear import of PNAs is from the cytoplasm is rapid in untreated cells and inhibited by NEM and GTPγS.

digitonin-permeabilized SK-BR-3 cells using confocal microscopy by adapting published methods.[5–9] We also used protection from restriction enzyme digestion within the intended PNA target sequence to detect PNA binding to the endogenous HER-2/neu target sequence by Southern hybridization as previously described.[3]

FIGURE 1 shows representative digital reconstruction images from confocal microscopy of digitonin-permeabilized SK-BR-3 cells. Fluorescently labeled, cationic, pyrimidine (cpy-) PNAs preferentially enter the nuclear compartment rather than remaining in the cytoplasm (FIG. 1 B). The nuclear uptake of this rhodamine-labeled, cpy-mono-PNA was significantly decreased by treating

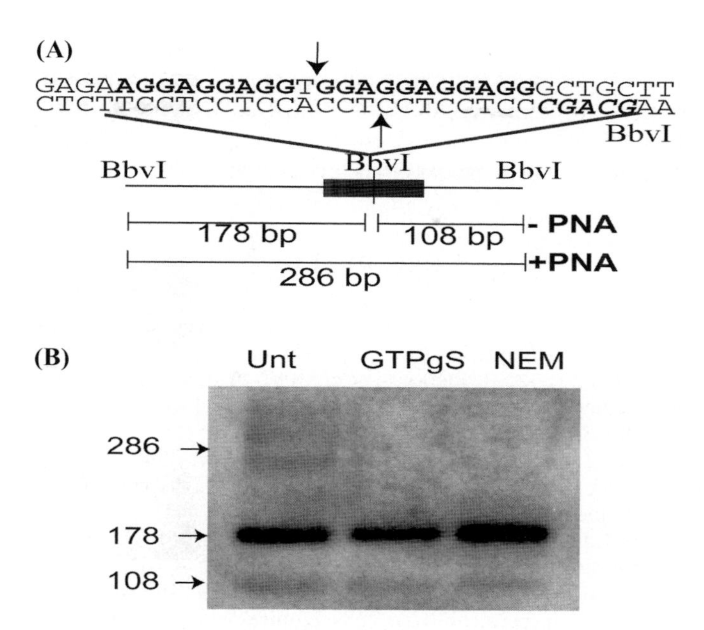

FIGURE 2. (**A**) PNA binding blocks a Bbv I site in the HER-2/neu promotor to generate a larger restriction fragments on Southern hybridization. (**B**) Digitonin-permeabilized SK-BR-3 cells were incubated with 10 μM bis-PNA (K-TJJTJJTJJ-OOO-CCTCCTCCT-KK) for 4 h in the absence or presence of nuclear transport inhibitors NEM or GTPγS. Protection was observed only in the absence of inhibitors.

the cells with NEM and GTPγS (FIG. 1 D and F). The binding of a cpy-bis-PNA to the HER-2/neu promotor was detectable after 2–4 h by restriction enzyme protection and Southern blot analysis (FIG. 2), but analysis after longer incubation times with the PNAs was not feasible with permeabilized cells. Treating the cells with NEM or GTPγS inhibited the binding of the PNA to its target sequence (FIG. 2 B), presumably by preventing the entry of the PNA into the nucleus. Cytoplasmic extracts improved nuclear uptake and target sequence binding in permeabilized cells, and we observed marked nuclear localization of PNAs after electroporation of intact cells (not shown). Collectively, these studies suggest that PNA uptake from the cytoplasm into the nucleus is an active process involving the NPC which is rapid, efficient, and does not require a NLS.

Molecules with a mass of less than 50 kDa could theoretically diffuse through the nuclear pores, but in most cases, active transport is involved in the uptake of large molecules.[4] Oligonucleotides have been reported to diffuse into the nucleus after microinjection into the cytoplasm.[10] The nuclear uptake of oligonucleotides probably depends on the length, backbone, and base composition of the oligomer. For example, energy-dependent, active nuclear transport of phosphodiester oligonucleotides through the NPC was reported for some sequences as a function of their base composition,[7] and the nuclear uptake of phosphorothioate oligonucleotides[8] and linear dsDNA[9] also involves active transport by the NPC. Our data suggest that the nuclear trafficking of PNAs is also an active and specific process that involves the NPC.

REFERENCES

1. KOPPELHUS, U. & P.E. NIELSEN. 2003. Cellular delivery of peptide nucleic acid (PNA). Adv. Drug Deliv. Rev. **55:** 267–280.
2. BRAUN, K., P. PESCHKE, R. PIPKORN, *et al.* 2002. A biological transporter for the delivery of peptide nucleic acids (PNAs) to the nuclear compartment of living cells. J. Mol. Biol. **318:** 237–243.
3. ZIEMBA, A.J., Z.V. ZHILINA, Y. KROTOVA-KHAN, *et al.* 2005. Targeting and regulation of the HER-2/neu oncogene promoter with bis-peptide nucleic acids. Oligonucleotides **15:** 36–50.
4. PEMBERTON, L.F. & B.M. PASCHAL. 2005. Mechanisms of receptor-mediated nuclear import and nuclear export. Traffic **6:** 187–198.
5. WILSON, G.L., B.S. DEAN, G. WANG & D.A. DEAN. 1999. Nuclear import of plasmid DNA in digitonin-permeabilized cells requires both cytoplasmic factors and specific DNA sequences. J. Biol. Chem. **274:** 22025–22032.
6. LIU, J., N. XIAO & D.B. DEFRANCO. 1999. Use of digitonin-permeabilized cells in studies of steroid receptor subnuclear trafficking. Methods **19:** 403–409.
7. HARTIG, R., R.L. SHOEMAN, A. JANETZKO, *et al.* 1998. Active nuclear import of single-stranded oligonucleotides and their complexes with non-karyophilic macromolecules. Biol. Cell **90:** 407–426.
8. LORENZ, P., T. MISTELI, B.F. BAKER, *et al.* 2000. Nucleocytoplasmic shuttling: a novel in vivo property of antisense phosphorothioate oligodeoxynucleotides. Nucleic Acids Res. **28:** 582–592.
9. SALMAN, H., D. ZBAIDA, Y. RABIN, *et al.* 2001. Kinetics and mechanism of DNA uptake into the cell nucleus. Proc. Natl. Acad. Sci. USA **98:** 7247–7252.
10. LEONETTI, J.P., N. MECHTI, G. DEGOLS, *et al.* 1991. Intracellular distribution of microinjected antisense oligonucleotides. Proc. Natl. Acad. Sci. USA **88:** 2702–2706.

TLR9 and the Recognition of Self and Non-Self Nucleic Acids

MARC S. LAMPHIER,[a] CHERILYN M. SIROIS,[b] ANJALI VERMA,[b] DOUGLAS T. GOLENBOCK,[b] AND EICKE LATZ[b]

[a]*Eisai Research Institute, Andover, Massachusetts 01810, USA*

[b]*Division of Infectious Diseases and Immunology, University of Massachusetts Medical School, Worcester, Massachusetts 01605, USA*

ABSTRACT: Toll-like receptors (TLRs) are involved in the innate recognition of foreign material and their activation leads to both innate and adaptive immune responses directed against invading pathogens. TLR9 is intracellularly expressed in the endo-lysosomal compartments of specialized immune cells. TLR9 is activated in response to DNA, in particular DNA containing unmethylated CpG motifs that are more prevalent in microbial than mammalian DNA. By detecting foreign DNA signatures TLR9 can sense the presence of certain viruses or bacteria inside the cell and mount an immune response. However, under certain conditions, TLR9 can also recognize self-DNA and this may promote immune pathologies with uncontrolled chronic inflammation. The autoimmune disease systemic lupus erythematosis (SLE) is characterized by the presence of immune stimulatory complexes containing autoantibodies against endogenous DNA and DNA- and RNA-associated proteins. Recent evidence indicates that the autoimmune response to these complexes involves TLR9 and the related single-stranded RNA-responsive TLRs 7 and 8, and therefore some breakdown in the normal ability of these TLRs to distinguish self and foreign DNA. Evidence suggests that immune cells use several mechanisms to discriminate between stimulatory and nonstimulatory DNA; however, it appears that TLR9 itself binds rather indiscriminately to a broad range of DNAs. We therefore propose that there is an additional recognition step by which TLR9 senses differences in the structures of bound DNA.

KEYWORDS: systemic lupus erythematosus (SLE); innate immune system; toll-like receptor (TLR); TLR9; TLR7; plasmacytpoid dendritic cell (pDC); B cell; Fc receptor; B cell receptor; innate immunity; interferon (IFN); CD80; CD86; ODN; oligonucleotides (ODN); immune complexes (IC); single-stranded DNA (ssDNA); double-stranded DNA (dsDNA); single-stranded RNA; double-stranded RNA

Address for correspondence: Eicke Latz, M.D., Ph.D., Division of Infectious Diseases and Immunology, University of Massachusetts Medical School, 364 Plantation Street, Worcester, MA 01605. Voice: 508-856-6554; fax: 508-856-5463.
e-mail: eicke.latz@umassmed.edu

**Ann. N.Y. Acad. Sci. 1082: 31–43 (2006). © 2006 New York Academy of Sciences.
doi: 10.1196/annals.1348.005**

THE MYSTERY IN COLEY'S TOXIN

The discovery of DNA as a potent immune stimulant has a long history that is intertwined with early attempts at cancer immunotherapy. In the 1890s, Dr. William Coley, a New York surgeon, investigated the use of various live and heat-killed bacteria for the treatment of tumors, with mixed efficacy.[1,2] Today, the attenuated mycobacterium bacillus Calmette-Guerin is approved as a pharmaceutical local treatment of superficial bladder cancer.[3–5] It is well recognized that microbial material contains a wide variety of immune stimulatory substances and, not surprisingly, the search for the active ingredient in Coley's bacterial lysates was extensive. In the 1980s, Tokunaga and associates elegantly identified bacterial DNA as the principal active component of Coley's bacterial lysates.[6] This laboratory went on to show in the early 1990s that the immune stimulatory effects were inherent in bacterial—but not vertebrate—DNA and that the effects on immune cells could be mimicked by short synthetic oligonucleotides (ODN).[7,8] At the same time it became apparent that synthetic single-stranded ODN designed as antisense nucleotides could have immune stimulatory properties.[9–11] In 1995, Krieg and colleagues showed that unmethylated CG dinucleotides (CpG) are responsible for most of the immune stimulatory properties of microbial DNA or synthetic ODN.[12,13] The cytosine bases in vertebrate DNA are generally methylated and CpG sequences are found less frequently than random utilization would predict (a phenomenon that is termed *CG suppression*).[14,15] Thus, immune cells appear to be able to discriminate and respond to fine structural differences between microbial and vertebrate DNA.

Both bacterial DNA and stimulatory CpG-containing ODN can strongly activate cells of the innate immune system, including macrophages and immature dendritic cells. Cellular activation by CpG-DNA leads to antigen presentation, upregulation of MHC class II molecules, and costimulatory molecules as well as release of proinflammatory cytokines and type I interferons (IFN).[12] The antitumor activities of the early Coley bacterial extracts may have been the result of the action of these cytokines, and there is renewed interest today in the use of defined immune stimulatory ODNs as immune modulators and anticancer therapies.[16]

TLR9 IS THE PRINCIPAL SIGNALING RECEPTOR FOR CPG-DNA

Toll-like receptors (TLRs) are a family of type I transmembrane receptors that are critically involved in the recognition of conserved molecular signatures that trigger the innate immune system.[17] At least 10 different TLRs are expressed in humans. TLRs are activated by a large number of conserved molecules associated with pathogens, suggesting that they function to alert the innate immune system to the presence of invading pathogens. However, TLRs can also be activated in some pathophysiological disease states that are associated with the presence of altered self-molecules.[18,19]

In 2000, it was shown that the receptor mediating the immune stimulatory effects of DNA is a member of the TLR family. By generating mice deficient in *tlr9* it was definitively demonstrated that TLR9 controls the cellular activation by CpG-DNA.[20] In the last decade, we have learned that TLR9 can recognize a variety of DNA sequences and that the quality and quantity of response can be dictated by the sequence and structure of DNA.[21–25] Over the last decade, several classes of stimulatory ODN based on their sequence and biological activities have been designed. Monomeric CpG-rich DNA, such as DNA sequences of the B-class (also known as D-class), can potently activate B cells and can induce maturation and production of inflammatory cytokines in plasmacytoid dendritic cells (pDCs).[22] These DNA sequences, however, lack the ability to induce type I IFNs in pDCs.[23] In contrast, A-class (also termed *K-class*) CpG-DNA can form secondary structures due to a central palindromic sequence and can assemble into nanoparticle-like complexes through the intermolecular interactions of polyG motifs that flank the central palindromic sequences.[24,26] This class of CpG-DNA can activate TLR9 to induce the production of large amounts of type I IFNs in pDCs. A third class of stimulatory CpG-DNA, the C-class CpG-DNA, combines the structural and functional features of A- and B-class. C-class ODN form hairpin structures as well as tertiary structures such as dimers and can lead to the production of type I IFNs.[21] It remains unclear how the formation of secondary structures in CpG-DNA can induce different signals from the same signaling receptor TLR9. It has been proposed that the IFN induction by A-class ODN depends on the duration and compartmentalization of ligand–receptor interaction in murine dendritic cell subtypes.[27] However, it is also conceivable that the structural CpG-DNA differences could lead to an alternative utilization of a co-receptor or a second signal from another receptor, which is required for IFN production. One should keep in mind that the different synthetic ODNs that are currently used to stimulate TLR9 have been developed to elicit a certain quantity of stimulation or quality of response. While particular CpG-DNA sequences are certainly helpful in elucidating the pharmacology of the TLR9 receptosome, the structures of naturally occurring TLR9 activators are most likely heterogeneous and differ from the synthetic CpG-DNA. For example, synthetic CpG-ODN contains phosphorothioate linkages in place of the phosphodiester linkages in naturally occurring DNA. Furthermore, the length and structure of naturally occurring DNA may differ depending on DNA source, status of enzymatic degradation, and the mode of DNA delivery. Therefore, natural microbial and endogenous DNA cannot be categorized into one of these classes of CpG-ODN and the cellular responses to DNA can differ greatly.

ENDOGENOUS DNA CAN LEAD TO INFLAMMATORY DISEASE VIA TLR9

Although recognition of DNA appears to serve to identify non-self pathogens, there is ample evidence that this ability to discriminate between foreign and self-DNA can sometimes be circumvented, leading to the development of autoimmune disease. The immune stimulatory property of DNA has recently been implicated in the pathogenesis of systemic lupus erythematosus (SLE). SLE pathogenesis involves loss of tolerance to self-antigens, which leads to the production of autoantibodies.[28,29] In sera of SLE patients autoantibodies are found against nuclear antigens, nucleic acids, and macromolecular complexes, such as chromatin or ribonucleoprotein particles. Tissue deposition of immune complexes (IC) can trigger immune-mediated tissue injury, which can lead to inflammation and injury of multiple organs.[30,31] ICs in SLE are believed to develop as part of a vicious cycle of an autoimmune response. The original trigger of autoimmunity remains elusive. However, it is clear that the disease is of a polygenic, multifactorial nature[32,33] and many of the genes that are defective in SLE patients are involved in the removal of cellular debris and nuclear material.[34–43] Thus, dysfunctional clearance of endogenous immunogenic material, such as may result from cell damage or viral infection, could trigger and/or maintain autoantibody production. The production of autoantibodies against DNA itself or against self-proteins associated with DNA results in DNA-containing ICs that are, as discussed below, capable of stimulating immune cells. SLE patients exhibit upregulation of IFN-induced genes,[44–46] normally the hallmark of viral infection, suggesting that continued stimulation by endogenous nucleic acids may be involved in the maintenance of the autoimmune state. Interestingly, the DNA that is recognized by antinuclear and anti-DNA antibodies in SLE patients has a CpG content that is similar to the composition of microbial DNA. The CpG frequency in SLE IC is 5 to 6 times higher than random DNA from the human genome and these CpG-rich complexes have been found to be immune stimulatory.[47,48]

For TLRs, the ability to selectively respond to self and non-self or altered self carries with it the inherent risk of the misinterpretation of molecular signatures. The pathogenesis of SLE may be one example of erroneous TLR activation, and studies in both mice and humans point to a role for TLR9 in this disease. Work originating from the Marshak–Rothstein group shows that rheumatoid factor autoreactive B cells can be activated by dual engagement of the antigen receptor and TLR9 (Fig. 1).[49] B cells that were non-reactive for rheumatoid factor could also sense IgG/DNA ICs, and this cellular activation required the presence of hypomethylated CpG motifs in the DNA.[50] In murine dendritic cells, chromatin-containing ICs activated cells in an Fc gamma receptor-dependent manner (Fig. 1).[51] These data imply that anti-DNA ICs can trigger systemic inflammation in a TLR9-dependent manner and suggest TLR9 as a potential target for pharmacologic intervention. Mouse *in vivo* studies, however, have

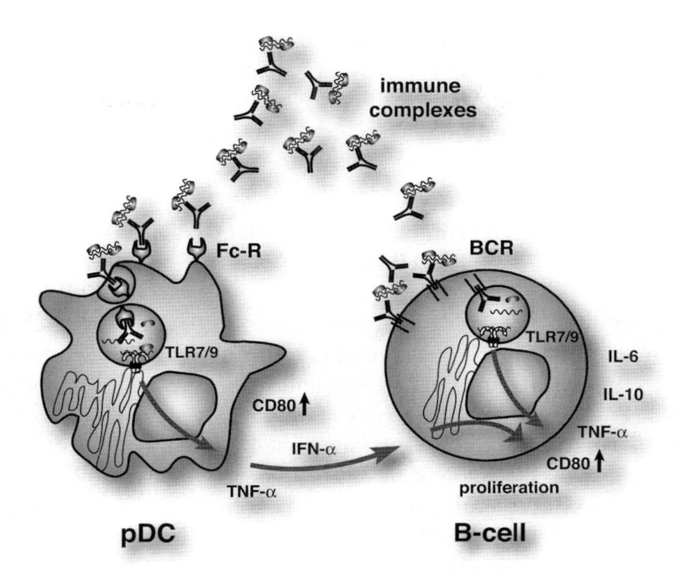

FIGURE 1. ICs containing DNA or RNA activate human pDCs and B cells via TLRs 7 and 9. DNA- or RNA-containing ICs isolated from patient sera or reconstituted *in vitro* can activate human pDCs and B cells. The uptake of ICs proceeds in pDCs mainly via Fc receptors and in B cells predominantly via the B cell receptor. Immune complex associated DNA or RNA can stimulate endosomally localized TLRs 7 or 9, which results in the activation of pDCs and B cells. PDCs are able to regulate B cell differentiation and antibody production as they can synergistically enhance B cell responses toward stimulatory ICs.

not produced a clear picture of the role of TLR9 in the initiation or progression of DNA-IC-associated diseases. While two studies have shown that lack of TLR9 in SLE-prone MRL/lpr mice resulted in inhibition of autoantibody production,[52,53] another study found unchanged autoantibody titers and increased immune pathology in mice of the same MRL/lpr background lacking *tlr9*.[54] This latter finding was partially confirmed in a second SLE-like mouse model involving *tlr9−/−* mice expressing a mutant form of phospholipase Cg2.[55] Yet, MRL/lpr mice deficient in TLR9 and TLR7 signaling due to knockout of the TLR adaptor molecule, MyD88, show decreased levels of chromatin-, Sm-, and rheumatoid factor-specific autoantibody titers.[56] Furthermore, this study also indicated that TLR7 recognizes RNA-containing IC and activates autoreactive B cells. Therefore, compensatory mechanisms between TLRs could explain the conflicting results regarding the *in vivo* role of TLR9 in IC-associated diseases. Thus, the role of TLRs 7 and 9 in disease should be tested together in autoimmune-prone mice.

In humans, IFN production by pDCs can be stimulated by small nuclear ribonucleoprotein particles, such as the prototypic autoantigen U1 snRNP, via TLR7 as well as by SLE IC via TLR9.[57,58] The immune stimulatory activity of circulating DNA- and RNA-containing IC requires uptake into intracellular compartments via Fc receptors. Thus, TLR9 recognition of DNA and TLR 7 and 8 recognition of RNA may both be important in human disease by mediating hyper-immune responses to DNA- or RNA-containing ICs (Fig. 1).

TLR9 RECOGNIZES DNA IN ENDO-LYSOSOMAL COMPARTMENTS

TLRs that primarily recognize microbial cell wall constituents, such TLRs 1, 2, 4, 5, and 6, are found on the surface of immune cells.[59–63] In contrast, TLRs 3, 7, 8, and 9 are localized to endosome and lysosome compartments inside the cell.[64–68] This group of TLRs presumably evolved to recognize nucleic acids of pathogens that have been endocytosed or phagocytosed, or viruses that have invaded the cell. Each of these TLRs is activated by different nucleic acid structures or related molecules: TLR3 is stimulated by double-stranded RNA,[69] TLRs 7 and 8 by single-stranded RNA and imidazoquinolines,[70–74] and TLR9 by single-stranded DNA.[20]

The mode of activation of these intracellular TLRs is complex and involves several indispensable steps. The first event in the recognition of immune stimulatory DNA by TLR9 is the uptake of DNA by the cells (FIG. 2). The uptake mechanism is also one of the least well-understood steps in this process. Uptake varies depending on the structure of the DNA. For example, many cell types can internalize single-stranded DNA. In contrast, double-stranded DNA enters most cells only poorly, although the uptake can be greatly enhanced by

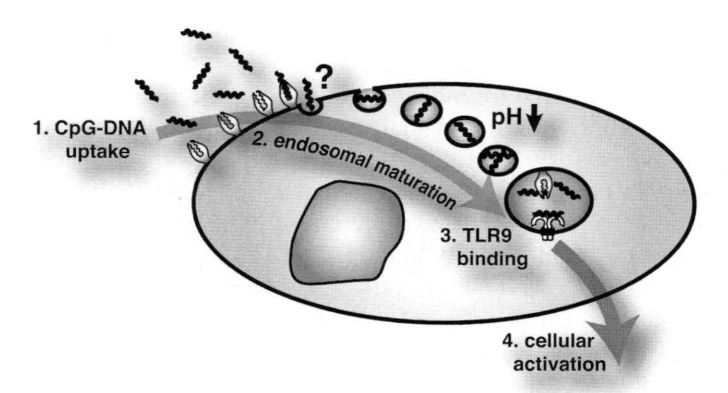

FIGURE 2. TLR9 activation proceeds in a multi-step process. Many cells can bind to and efficiently take up single-stranded DNA in a receptor-mediated pathway. The surface receptor has not been clearly identified (1). In a process termed *endosomal maturation* (2) endosomes are acidified and binding to TLR9 takes place (3). The binding of CpG-TLR9 results in the initiation of signaling ultimately leading to various aspects of cellular activation (4).

complexing DNA with cationic lipids. Similarly, most cells do not efficiently take up single-stranded RNA, but cationic lipids can also facilitate uptake, as well as stimulation of TLRs 7 and 8. Double-stranded RNA enters cells with modest efficiency, and its efficacy is also enhanced by cationic lipids or, more classically, by mixture with DEAE-dextran. A number of studies have examined uptake of single-stranded DNA, particularly in the context of antisense nucleotide research. Single-stranded DNA appears to enter cells by receptor-mediated endocytosis.[75–79] A number of cell-surface proteins have been described that may mediate uptake of single-stranded DNA including scavenger receptors,[80] CD11b/CD18,[81] and a variety of uncharacterized binding proteins.[75,78,82,83] However, conclusive evidence for a particular single-stranded DNA receptor has yet to be demonstrated (FIG. 2). Uptake can be bypassed by the use of cationic lipids, and many nonstimulatory DNAs become stimulatory when introduced via this method, suggesting that selectivity is in part determined by uptake.[84]

The subsequent step in signaling to nucleic acids, endosomal maturation, is also an imprecisely defined process (FIG. 2). Drugs that interfere with endosomal acidification, such as bafilomycin, can block signaling induced by nucleic acids or homologous molecules, suggesting that a low pH environment is required for signaling.[85] The antimalarial drug, chloroquine, can raise the pH in the endosomal microenvironment and is also a potent TLR9 inhibitor. However, early studies on chloroquine have also demonstrated that it can bind DNA and thereby could theoretically also impair the recognition of stimulatory DNA.[86] There is experimental evidence that DNA binding to TLR9 has an optimal pH between 5 and 6.[87] This finding suggests that there are regulatory mechanisms that only allow TLR9 activation in certain pH-defined subsets of endosomes and prevent activation of TLR9 in other subsets or outside endosomes. Additionally, DNA is likely to be highly concentrated within the endosome, as is the case for other endosomally targeted molecules, and the concentration of DNA could thus influence how and which DNA sequences can activate TLR9.[84] The significance of the endosomal location of DNA recognition is underlined by a study showing that TLR9 engineered to localize to the cell surface can respond to self-DNA and fails to respond to viral DNA.[88] Similarly, immune complex uptake is facilitated by the B cell receptor and/or Fc receptors and IC-associated nucleic acids are only stimulatory after uptake.[57,58]

MECHANISM OF TLR9 ACTIVATION

The molecular mechanisms leading to the activation TLR9 by DNA are not fully understood. Several studies and methodologies have indicated that TLR9 can directly bind CpG-DNA. The consensus now emerging is that DNA binding to TLR9 does not seem to determine functional outcome, since both stimulatory and nonstimulatory DNA sequences are shown to bind TLR9.[65,89] These findings are supported by the fact that nonstimulatory (control) CpG-DNA can competitively block TLR9 activation by stimulatory DNA sequences and compete for binding at the level of TLR9.[65] Potent TLR9-inhibiting sequences, so-called inhibitory ODN, have been described. These so-called inhibitory ODN are approximately 15 bases long, have a phosphorothioate backbone, and contain two stretches of sequences that influence their inhibitory function: a pyrimidine-rich triplet, preferably CCT, which is positioned with a spacer 5' to a GGG sequence.[90–92] It is likely that these oligonucleotides act as direct TLR9 antagonists that can efficiently compete with stimulatory DNA for binding, yet not trigger signaling. Along this line, our own unpublished work using a homogenous binding assay indicates that crude sequence specificity can be observed when comparing homopolymers. For example, poly-adenine (poly-A) shows low affinity binding whereas poly-T binds with higher affinity. An important criterion for both binding to and signaling via TLR seems to be the length of the ODN and/or the degree to which the ODN are clustered. Collectively, these results suggest that differences in ligand binding may not entirely explain the discrimination between stimulatory and nonstimulatory DNAs. Rather, TLR9 may bind to a broad range of nucleic acids, but only respond to a subset of these, suggesting a signaling event that is subsequent to, and distinct from, DNA binding. This event could be a receptor rearrangement after binding to stimulatory sequences or the recruitment of a cofactor that associates with TLR9 and provides a sequence-specific signaling function. Although a cofactor that mediates sequence-specific signaling cannot be ruled out, a more compelling explanation is that TLR9 itself possesses both a sequence non-specific binding function and a motif-specific signaling function. For example, there may be multiple sites of DNA interaction on TLR9, as was suggested for TLR3 and double-stranded RNA.[93] In this latter scenario, stimulatory ODN would coordinate and/or rotate the extracellular domain(s) of TLR9 in a way that orients the signaling domains in a signaling-competent manner. Together, our understanding of the molecular events in TLR9 function is still fragmentary and hypothetical. It remains of pivotal importance to define the exact molecular mechanisms of TLR9 activation, as this will facilitate the development of TLR9-modulating therapies for a broad range of immune pathologies.

ACKNOWLEDGMENT

This work was funded by grants of the National Institute of Health.

REFERENCES

1. Coley, W.B. 1991. The treatment of malignant tumors by repeated inoculations of erysipelas. With a report of ten original cases. 1893. Clin. Orthop. Relat. Res. **263:** 3–11.

2. WIEMANN, B. & C.O. STARNES. 1994. Coley's toxins, tumor necrosis factor and cancer research: a historical perspective. Pharmacol Ther. **64:** 529–564.
3. BASSI, P. 2002. BCG (Bacillus of Calmette Guerin) therapy of high-risk superficial bladder cancer. Surg Oncol. **11:** 77–83.
4. EIDINGER, D. & A. MORALES. 1978. BCG immunotherapy of metastatic adenocarcinoma of the kidney. Natl .Cancer Inst. Monogr. **49:** 339–341.
5. MORALES, A. 1978. Adjuvant immunotherapy in superficial bladder cancer. Natl. Cancer Inst. Monogr. **49:** 315–319.
6. TOKUNAGA, T. *et al.* 1984. Antitumor activity of deoxyribonucleic acid fraction from *Mycobacterium bovis* BCG. I. Isolation, physicochemical characterization, and antitumor activity. J. Natl. Cancer Inst. **72:** 955–962.
7. YAMAMOTO, S. *et al.* 1992. DNA from bacteria, but not from vertebrates, induces interferons, activates natural killer cells and inhibits tumor growth. Microbiol. Immunol. **36:** 983–897.
8. YAMAMOTO, T. *et al.* 1994. Synthetic oligonucleotides with certain palindromes stimulate interferon production of human peripheral blood lymphocytes *in vitro*. Jpn. J. Cancer Res. **85:** 775–779.
9. BRANDA, R.F. *et al.* 1993. Immune stimulation by an antisense oligomer complementary to the rev gene of HIV-1. Biochem. Pharmacol. **45:** 2037–2043.
10. HARTMANN, G. *et al.* 1996. Oligodeoxynucleotides enhance lipopolysaccharide-stimulated synthesis of tumor necrosis factor: dependence on phosphorothioate modification and reversal by heparin. Mol. Med. **2:** 429–438.
11. TANAKA, T., C.C. CHU & W.E. PAUL. 1992. An antisense oligonucleotide complementary to a sequence in I gamma 2b increases gamma 2b germline transcripts, stimulates B cell DNA synthesis, and inhibits immunoglobulin secretion. J. Exp. Med. **175:** 597–607.
12. KRIEG, A.M. 2002. CpG motifs in bacterial DNA and their immune effects. Annu. Rev. Immunol. **20:** 709–760.
13. KRIEG, A.M. *et al.* 1995. CpG motifs in bacterial DNA trigger direct B-cell activation. Nature **374:** 546–549.
14. AISSANI, B. & G. BERNARDI. 1991. CpG islands: features and distribution in the genomes of vertebrates. Gene. **106:** 173–183.
15. ANTEQUERA, F. & A. BIRD. 1993. CpG islands. *In* DNA Methylation: molecular Biology and Biological Significance. P.J. Jost & P.H. Saluz, Eds.: 169–185. Birkhauser Verlag. Basel, Switzerland.
16. KRIEG, A.M. 2006. Therapeutic potential of Toll-like receptor 9 activation. Nat. Rev. Drug Discov. **5:** 471–484.
17. JANEWAY, C.A., JR. 1992. The immune system evolved to discriminate infectious nonself from noninfectious self. Immunol. Today. **13:** 11–16.
18. TAKEDA, K., T. KAISHO & S. AKIRA. 2003. Toll-like receptors. Annu. Rev. Immunol. **21:** 335–376.
19. MATZINGER, P. 1994. Tolerance, danger, and the extended family. Annu. Rev. Immunol. **12:** 991–1045.
20. HEMMI, H. *et al.* 2000. A Toll-like receptor recognizes bacterial DNA. Nature **408:** 740–745.
21. HARTMANN, G. *et al.* 2003. Rational design of new CpG oligonucleotides that combine B cell activation with high IFN-alpha induction in plasmacytoid dendritic cells. Eur. J. Immunol. **33:** 1633–1641.
22. HARTMANN, G. & A.M. KRIEG. 2000. Mechanism and function of a newly identified CpG DNA motif in human primary B cells. J Immunol. **164:** 944–953.

23. KRUG, A. *et al.* 2001. Identification of CpG oligonucleotide sequences with high induction of IFN-alpha/beta in plasmacytoid dendritic cells. Eur. J. Immunol. **31:** 2154–2163.
24. MARSHALL, J.D. *et al.* 2003. Identification of a novel CpG DNA class and motif that optimally stimulate B cell and plasmacytoid dendritic cell functions. J. Leukoc. Biol. **73:** 781–792.
25. VERTHELYI, D. *et al.* 2001. Human peripheral blood cells differentially recognize and respond to two distinct CPG motifs. J. Immunol. **166:** 2372–2377.
26. KERKMANN, M. *et al.* 2005. Spontaneous formation of nucleic acid-based nanoparticles is responsible for high interferon-alpha induction by CpG-A in plasmacytoid dendritic cells. J. Biol. Chem. **280:** 8086–8093.
27. HONDA, K. *et al.* 2005. Spatiotemporal regulation of MyD88-IRF-7 signalling for robust type-I interferon induction. Nature **434:** 1035–1040.
28. LAWRENCE, R.C. *et al.* 1998. Estimates of the prevalence of arthritis and selected musculoskeletal disorders in the United States. Arthritis Rheum. **41:** 778–799.
29. MOK, C.C. & C.S. LAU. 2003. Pathogenesis of systemic lupus erythematosus. J. Clin. Pathol. **56:** 481–490.
30. TAN, E.M. *et al.* 1982. The 1982 revised criteria for the classification of systemic lupus erythematosus. Arthritis Rheum. **25:** 1271–1277.
31. HOCHBERG, M.C. 1997. Updating the American College of Rheumatology revised criteria for the classification of systemic lupus erythematosus. Arthritis Rheum. **40:** 1725.
32. TSAO, B.P. 2003. The genetics of human systemic lupus erythematosus. Trends Immunol. **24:** 595–602.
33. TSAO, B.P. 2004. Update on human systemic lupus erythematosus genetics. Curr. Opin. Rheumatol. **16:** 513–521.
34. BATTEUX, F. *et al.* 1999. FCgammaRII (CD32)-dependent induction of interferon-alpha by serum from patients with lupus erythematosus. Eur. Cytokine Netw. **10:** 509–514.
35. RAVETCH, J.V. & R.A. CLYNES. 1998. Divergent roles for Fc receptors and complement *in vivo*. Annu. Rev. Immunol. **16:** 421–432.
36. HANAYAMA, R. *et al.* 2004. Autoimmune disease and impaired uptake of apoptotic cells in MFG-E8-deficient mice. Science **304:** 1147–1150.
37. SHAW, P.X. *et al.* 2000. Natural antibodies with the T15 idiotype may act in atherosclerosis, apoptotic clearance, and protective immunity. J. Clin. Invest. **105:** 1731–1740.
38. KIM, S.J. *et al.* 2002. I-PLA(2) activation during apoptosis promotes the exposure of membrane lysophosphatidylcholine leading to binding by natural immunoglobulin M antibodies and complement activation. J. Exp. Med. **196:** 655–665.
39. BOTTO, M. *et al.* 1998. Homozygous C1q deficiency causes glomerulonephritis associated with multiple apoptotic bodies. Nat. Genet. **19:** 56–59.
40. LU, Q. & G. LEMKE. 2001. Homeostatic regulation of the immune system by receptor tyrosine kinases of the Tyro 3 family. Science **293:** 306–311.
41. KABAROWSKI, J.H. *et al.* 2001. Lysophosphatidylcholine as a ligand for the immunoregulatory receptor G2A. Science **293:** 702–705.
42. BICKERSTAFF, M.C. *et al.* 1999. Serum amyloid P component controls chromatin degradation and prevents antinuclear autoimmunity. Nat. Med. **5:** 694–647.
43. NAPIREI, M. *et al.* 2000. Features of systemic lupus erythematosus in Dnase1-deficient mice. Nat. Genet. **25:** 177–181.

44. Bengtsson, A.A. *et al.* 2000. Activation of type I interferon system in systemic lupus erythematosus correlates with disease activity but not with antiretroviral antibodies. Lupus **9:** 664–671.
45. Crow, M.K., K.A. Kirou & J. Wohlgemuth. 2003. Microarray analysis of interferon-regulated genes in SLE. Autoimmunity **36:** 481–490.
46. Hooks, J.J. *et al.* 1979. Immune interferon in the circulation of patients with autoimmune disease. N. Engl. J. Med. **301:** 5–8.
47. Sano, H. *et al.* 1989. Binding properties of human anti-DNA antibodies to cloned human DNA fragments. Scand. J. Immunol. **30:** 51–63.
48. Sato, Y. *et al.* 1999. CpG motif-containing DNA fragments from sera of patients with systemic lupus erythematosus proliferate mononuclear cells *in vitro*. J. Rheumatol. **26:** 294–301.
49. Leadbetter, E.A. *et al.* 2002. Chromatin-IgG complexes activate B cells by dual engagement of IgM and Toll-like receptors. Nature **416:** 603–607.
50. Viglianti, G.A. *et al.* 2003. Activation of autoreactive B cells by CpG dsDNA. Immunity **19:** 837–847.
51. Boule, M.W. *et al.* 2004. Toll-like receptor 9-dependent and -independent dendritic cell activation by chromatin-immunoglobulin G complexes. J. Exp. Med. **199:** 1631–1640.
52. Christensen, S.R. *et al.* 2005. Toll-like receptor 9 controls anti-DNA autoantibody production in murine lupus. J. Exp. Med. **202:** 321–331.
53. Lartigue, A. *et al.* 2006. Role of TLR9 in anti-nucleosome and anti-DNA antibody production in lpr mutation-induced murine lupus. J. Immunol. **177:** 1349–1354.
54. Wu, X. & S.L. Peng. 2006. Toll-like receptor 9 signaling protects against murine lupus. Arthritis Rheum. **54:** 336–342.
55. Yu, P. *et al.* 2006. Toll-like receptor 9-independent aggravation of glomerulonephritis in a novel model of SLE. Int. Immunol. **18(8):** 1211–1219.
56. Lau, C.M. *et al.* 2005. RNA-associated autoantigens activate B cells by combined B cell antigen receptor/Toll-like receptor 7 engagement. J. Exp. Med. **202:** 1171–1177.
57. Vollmer, J. *et al.* 2005. Immune stimulation mediated by autoantigen binding sites within small nuclear RNAs involves Toll-like receptors 7 and 8. J. Exp. Med. **202:** 1575–1585.
58. Means, T.K. *et al.* 2005. Human lupus autoantibody-DNA complexes activate DCs through cooperation of CD32 and TLR9. J. Clin. Invest. **115:** 407–417.
59. Latz, E. *et al.* 2002. Lipopolysaccharide rapidly traffics to and from the Golgi apparatus with the toll-like receptor 4-MD-2-CD14 complex in a process that is distinct from the initiation of signal transduction. J. Biol. Chem. **277:** 47834–47843.
60. Latz, E. *et al.* 2003. The LPS receptor generates inflammatory signals from the cell surface. J. Endotoxin. Res. **9:** 375–380.
61. Husebye, H. *et al.* 2006. Endocytic pathways regulate Toll-like receptor 4 signaling and link innate and adaptive immunity. Embo J. **25:** 683–692.
62. Sandor, F. *et al.* 2003. Importance of extra- and intracellular domains of TLR1 and TLR2 in NFkappa B signaling. J. Cell Biol. **162:** 1099–1110.
63. Visintin, A. *et al.* 2003. Lysines 128 and 132 enable lipopolysaccharide binding to MD-2, leading to Toll-like receptor-4 aggregation and signal transduction. J. Biol. Chem. **278:** 48313–48320.

64. AHMAD-NEJAD, P. *et al.* 2002. Bacterial CpG-DNA and lipopolysaccharides activate Toll-like receptors at distinct cellular compartments. Eur. J. Immunol. **32:** 1958–1968.
65. LATZ, E. *et al.* 2004. TLR9 signals after translocating from the ER to CpG DNA in the lysosome. Nat. Immunol. **5:** 190–198.
66. LEIFER, C.A. *et al.* 2004. TLR9 is localized in the endoplasmic reticulum prior to stimulation. J. Immunol. **173:** 1179–1183.
67. LATZ, E. *et al.* 2004. Mechanisms of TLR9 activation. J. Endotoxin. Res. **10:** 406–412.
68. ESPEVIK, T. *et al.* 2003. Cell distributions and functions of Toll-like receptor 4 studied by fluorescent gene constructs. Scand. J. Infect. Dis. **35:** 660–664.
69. ALEXOPOULOU, L. *et al.* 2001. Recognition of double-stranded RNA and activation of NF-kappaB by Toll-like receptor 3. Nature **413:** 732–738.
70. LUND, J.M. *et al.* 2004. Recognition of single-stranded RNA viruses by Toll-like receptor 7. Proc. Natl. Acad. Sci. USA **101:** 5598–5603.
71. HEIL, F. *et al.* 2004. Species-specific recognition of single-stranded RNA via toll-like receptor 7 and 8. Science **303:** 1526–1529.
72. DIEBOLD, S.S. *et al.* 2004. Innate antiviral responses by means of TLR7-mediated recognition of single-stranded RNA. Science **303:** 1529–1531.
73. HEMMI, H. *et al.* 2002. Small anti-viral compounds activate immune cells via the TLR7 MyD88-dependent signaling pathway. Nat. Immunol. **3:** 196–200.
74. HEIL, F. *et al.* 2003. The Toll-like receptor 7 (TLR7)-specific stimulus loxoribine uncovers a strong relationship within the TLR7, 8 and 9 subfamily. Eur. J. Immunol. **33:** 2987–2997.
75. LOKE, S.L. *et al.* 1989. Characterization of oligonucleotide transport into living cells. Proc. Natl. Acad. Sci. USA **86:** 3474–3478.
76. YAKUBOV, L.A. *et al.* 1989. Mechanism of oligonucleotide uptake by cells: involvement of specific receptors? Proc. Natl. Acad. Sci. USA **86:** 6454–6458.
77. DE DIESBACH, P. *et al.* 2002. Receptor-mediated endocytosis of phosphodiester oligonucleotides in the HepG2 cell line: evidence for non-conventional intracellular trafficking. Nucleic Acids Res. **30:** 1512–1521.
78. DE DIESBACH, P. *et al.* 2000. Identification, purification and partial characterisation of an oligonucleotide receptor in membranes of HepG2 cells. Nucleic Acids Res. **28:** 868–874.
79. NAKAI, D. *et al.* 1996. Cellular uptake mechanism for oligonucleotides: involvement of endocytosis in the uptake of phosphodiester oligonucleotides by a human colorectal adenocarcinoma cell line, HCT-15. J. Pharmacol. Exp. Ther. **278:** 1362–1372.
80. ZHU, F.G., C.F. REICH & D.S. PISETSKY. 2001. The role of the macrophage scavenger receptor in immune stimulation by bacterial DNA and synthetic oligonucleotides. Immunology **103:** 226–234.
81. BENIMETSKAYA, L. *et al.* 1997. Mac-1 (CD11b/CD18) is an oligodeoxynucleotide-binding protein. Nat. Med. **3:** 414–420.
82. LAKTIONOV, P.P. *et al.* 1999. Characterisation of membrane oligonucleotide-binding proteins and oligonucleotide uptake in keratinocytes. Nucleic Acids Res. **27:** 2315–2324.
83. YAO, G.Q., S. CORRIAS & Y.C. CHENG. 1996. Identification of two oligodeoxyribonucleotide binding proteins on plasma membranes of human cell lines. Biochem. Pharmacol. **51:** 431–436.

84. YASUDA, K. *et al.* 2005. Endosomal translocation of vertebrate DNA activates dendritic cells via TLR9-dependent and -independent pathways. J. Immunol. **174:** 6129–6136.
85. HACKER, H. *et al.* 1998. CpG-DNA-specific activation of antigen-presenting cells requires stress kinase activity and is preceded by non-specific endocytosis and endosomal maturation. Embo J. **17:** 6230–6240.
86. ALLISON, J.L., R. L. O'BRIEN & F.E. HAHN. 1965. DNA: reaction with chloroquine. Science **149:** 1111–1113.
87. RUTZ, M. *et al.* 2004. Toll-like receptor 9 binds single-stranded CpG-DNA in a sequence- and pH-dependent manner. Eur. J. Immunol. **34:** 2541–2550.
88. BARTON, G.M., J.C. KAGAN & R. MEDZHITOV. 2006. Intracellular localization of Toll-like receptor 9 prevents recognition of self DNA but facilitates access to viral DNA. Nat. Immunol. **7:** 49–56.
89. YASUDA, K. *et al.* 2006. CpG motif-independent activation of TLR9 upon endosomal translocation of "natural" phosphodiester DNA. Eur. J. Immunol. **36:** 431–436.
90. KLINMAN, D.M. *et al.* 2003. Regulation of CpG-induced immune activation by suppressive oligodeoxynucleotides. Ann. N. Y. Acad. Sci. **1002:** 112–123.
91. LENERT, P. 2005. Inhibitory oligodeoxynucleotides – therapeutic promise for systemic autoimmune diseases? Clin. Exp. Immunol. **140:** 1–10.
92. ASHMAN, R.F. *et al.* 2005. Sequence requirements for oligodeoxyribonucleotide inhibitory activity. Int. Immunol. **17:** 411–420.
93. CHOE, J., M.S. KELKER & I.A. WILSON. 2005. Crystal structure of human toll-like receptor 3 (TLR3) ectodomain. Science **309:** 581–585.

RNA Silencing in the Struggle against Disease

VOLKER PATZEL, ISABELL DIETRICH, AND STEFAN H.E. KAUFMANN

Department of Immunology, Max-Planck-Institute for Infection Biology, D–10117 Berlin, Germany

ABSTRACT: Numerous acquired and hereditary diseases are caused by aberrant cellular or microbial gene expression. As a result of sequencing of the human genome and the genomes of various human pathogens, researchers have gained access to a large number of genes with residual functions. For functional validation of unknown genes, their functions can be specifically inhibited by antisense nucleic acids or small interfering RNAs (siRNAs) and the consequences of the functional loss, that is, the resulting phenotypes, can be analyzed. While antisense nucleic acids block the translation stoichiometrically by docking on the mRNA, siRNAs induce a highly effective cellular mechanism that causes catalytic destruction of several mRNA molecules by a single siRNA molecule. This mechanism, called RNA interference (RNAi), is only intrinsic to eukaryotic cells. Consequently, only eukaryotic target validation is pushed by RNAi whereas time-consuming conventional knockout techniques or the less efficient antisense strategies have to be applied for prokaryotic target validation. We succeeded in triggering gene silencing by siRNA in prokaryotic cells. This opens promising perspectives regarding validation of prokaryotic gene functions.

KEYWORDS: gene silencing; siRNAs; RNAi; prokaryotic; genome; prokaryotes; bacteria

Macroorganisms, that is, eukaryotes, make use of complex molecular and/or cellular systems to combat microbial attacks. Microorganisms, such as prokaryotes or viruses, have evolved sophisticated stratagems in order to counteract or escape from host defense and succeed in long-term survival by misuse of the infected macroorganisms. Thus, microbial infection relies on a variety of gene products originating from both sides, the infectious microorganism and the infected macroorganism. In the battle against infectious diseases, RNA technologies make vital contributions toward identification and validation of genes involved in host–pathogen interactions. Such genes, that is, their mRNAs

Address for correspondence: Volker Patzel, Department of Immunology, Max-Planck-Institute for Infection Biology, Schumannstrasse 21/22, D–10117 Berlin, Germany. Voice: +49-0-30-28460520; fax: +49-0-30-28460505.

e-mail: patzel@mpiib-berlin.mpg.de

Ann. N.Y. Acad. Sci. 1082: 44–46 (2006). © 2006 New York Academy of Sciences. doi: 10.1196/annals.1348.051

and gene products, represent important diagnostic and therapeutic target structures. This has been made possible by the full decryption of the human genome and many microbial genomes. Within the scope of so-called transcriptome and proteome analyses, eukaryotic and prokaryotic drug target candidate genes can be identified; however, functions can only be assigned to a small fraction of these genes. The discovery that small interfering RNAs (siRNAs) can specifically trigger RNA interference (RNAi) in mammalian cells thereby bypassing unspecific interferon responses has led to a renaissance of eukaryotic target validation.[1] RNAi has made it possible to perform functional validation of eukaryotic targets in largely automated high-throughput (HT) processes. Conversely, the RNAi machinery or crucial components thereof, are not intrinsic to prokaryotic cells, although, analogs to eukaryotic Argonaute proteins, which represent the core of the RNA-induced silencing complex (RISC) and RISC-related complexes,[2–8] have been described in prokaryotes, as well.[9] Thus, prokaryotic target validation was still restricted to HT-incompatible cost- and labor-intensive knockout strategies or less efficient antisense technologies. To overcome this bottleneck in prokaryotic target validation we investigated whether it would be possible to complement the prokaryotic repertoire by eukaryotic functions in order to create a functional silencing machinery for use of RNAi in prokaryotes. Based on this strategy we succeeded in silencing

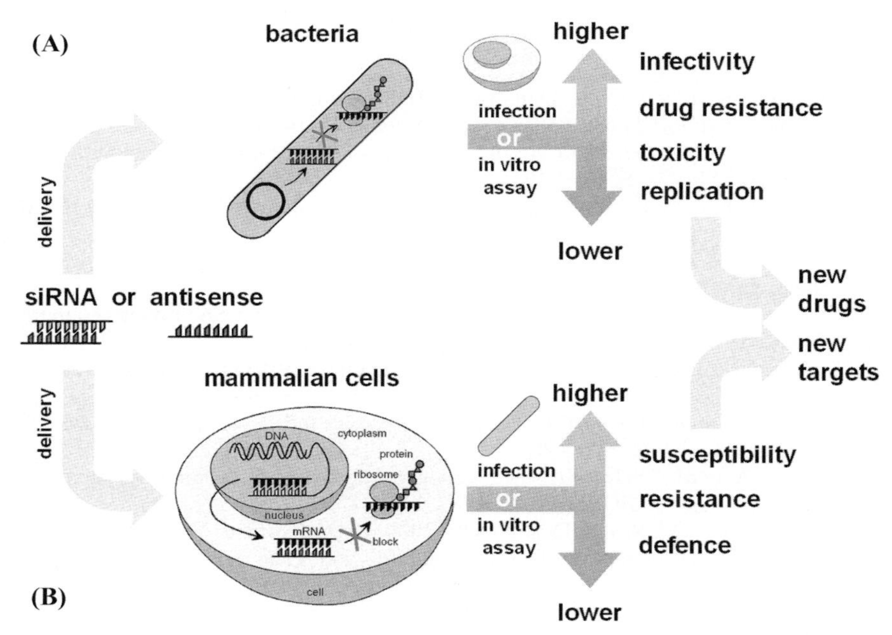

FIGURE 1. Prokaryotic RNA silencing (**A**) completes conventional eukaryotic (**B**) target validation and drug development form the pathogen side.

prokaryotic genes in Gram-positive and Gram-negative bacteria as well as in Mycobacteria. Interestingly, observed phenotypes ranged from transient gene knockdown to persistent gene knockout. From this point of view, siRNA-triggered prokaryotic gene silencing proved even more powerful compared to eukaryotic RNAi. Gene knockdown/out bacteria can be monitored for replication, infectivity, toxicity, and drug resistance *in vitro*, using HT technologies, and *in vivo* thereby efficiently completing functional target validation from the pathogen side (Fig. 1). As ultimate goal of these efforts we envisage the creation of a global landscape of genes and functions for infection and defence against microbial pathogens. Such a profile can then serve as basis for (*a*) the creation of a diagnostic tool for detection of individuals with increased susceptibility to bacterial infectious agents and (*b*) the identification of genes which are relevant for control of microbial of pathogens. These genes and their products represent highly promising target structures for vaccine and drug development.

ACKNOWLEDGMENTS

This work was supported by a grant from the Fonds Chemie and the RNA Network that is funded by the EFRE, the BMBF, and Chiron Corp.

REFERENCES

1. ELBASHIR, S.M. *et al.* 2001. Duplexes of 21-nucleotide RNAs mediate RNA interference in cultured mammalian cells. Nature **411:** 494–498.
2. HAMMOND, S.M. *et al.* 2001. Argonaute2, a link between genetic and biochemical analyses of RNAi. Science **293:** 1146–1150.
3. HUTVÁGNER, G. & P.Z. ZAMORE. 2002. A microRNA in a multiple-turnover RNAi enzyme complex. Science **297:** 2056–2060.
4. MARTINEZ, J. *et al.* 2002. Single-stranded antisense siRNAs guide target RNA cleavage in RNAi. Cell **110:** 563–574.
5. MOURELATOS, Z. *et al.* 2002. miRNPs: a novel class of ribonucleoproteins containing numerous microRNAs. Genes Dev. **16:** 720–728.
6. LIU, J. *et al.* 2004. Argonaute2 is the catalytic engine of mammalian RNAi. Science **305:** 1437–1441.
7. TOMARI, Y. *et al.* 2004. RISC assembly defects in the Drosophila RNAi mutant armitage. Cell **116:** 831–841.
8. VERDEL, A. *et al.* 2004. RNAi-mediated targeting of heterochromatin by the RITS complex. Science **303:** 672–676.
9. PARKER, J.S., S.M. ROE & D. BARFORD. 2004. Crystal structure of a PIWI protein suggests mechanisms for siRNA recognition and slicer activity. EMBO J. **23:** 4727–4737.

Cellular Dynamics of Antisense Oligonucleotides and Short Interfering RNAs

LI KIM LEE,[a] BRANDEE M. DUNHAM,[a] ZHUTING LI,[b] AND CHARLES M. ROTH[a,b]

[a]*Department of Chemical and Biochemical Engineering, Rutgers University, Piscataway, New Jersey 08854-8058, USA*

[b]*Department of Biomedical Engineering, Rutgers University, Piscataway, New Jersey 08854-8058, USA*

ABSTRACT: We aim to compare quantitatively the dynamics of the effectiveness of antisense oligonucleotides (AS ODNs) versus short interfering RNAs (siRNAs) and relate their effectiveness to sequence metrics (e.g., predicted free energy of binding). AS ODNs against a quantitative model target, pd1EGFP (destabilized enhanced GFP [green fluorescent protein]), were selected using our thermodynamic model, and siRNA sequences were designed to be identical to the AS ODN sequences in the antisense strand. We evaluated d1EGFP inhibition in transiently and stably transfected Chinese hamster ovary (CHO) cells over time using flow cytometry. Overall, our results show that the rationally designed AS ODN and siRNA sequences proved effective inhibitors of GFP expression and suggest that certain regions of mRNA may be susceptible to both AS ODNs and siRNAs.

KEYWORDS: green fluorescent protein; thermodynamic model; sequence selection; transient transfection

INTRODUCTION

Antisense technology and, more recently, RNA interference (RNAi) have rapidly developed into significant tools for determining gene function and validating small molecule drug targets, especially in cell culture systems. Major challenges associated with the development of antisense oligonucleotides (AS ODNs) and short interfering RNAs (siRNAs) as viable therapeutic agents

Address for correspondence: Charles M. Roth, Department of Chemical and Biochemical Engineering, Rutgers University, 98 Brett Road C-228, Piscataway, NJ 08854-8058. Voice: +732-445-4109; fax: +732-445-2581.
e-mail: cmroth@rci.rutgers.edu

Ann. N.Y. Acad. Sci. 1082: 47–51 (2006).
doi: 10.1196/annals.1348.061

include potency and specificity. Effective inhibitors should possess high binding affinity to the target RNA sequence and hybridize efficiently to the mRNA target without nonspecific binding.

Higher order structures of mRNA can have considerable impact on the hybridization efficiency and potency, and correspondingly, the activity, of AS ODNs *in vitro*. To address this, we previously developed a thermodynamic model[1,2] that accounts for the energetics of structural alterations in both the target mRNA and the ODN. Using the Oligowalk utility of RNAStructure 4.11 (available at http://rna.urmc.rochester.edu), we identified ODN sequences with the highest predicted binding affinity for our target mRNA of interest, d1EGFP, a destabilized variant of the enhanced green fluorescent protein (GFP). The short half-life (1 h) of the protein allowed for a close coupling between inhibition of translation and decrease in GFP signal.

Three thermodynamically selected AS sequences (ODN 157, ODN 703, and ODN 708) were experimentally compared with siRNAs chosen to have consistent sequences, that is, targeted to the same region on the mRNA. For reference, we also compared these sequences with one AS ODN (ODN 616)[3] and one siRNA (siRNA 218)[4] from the literature. The literature sequences were predicted by our model (using the corresponding antisense strand for the published siRNA) to be less effective inhibitors.

RESULTS

Our experimental strategy involved delivery of AS ODNs or siRNAs to either a Chinese hamster ovary (CHO) cell line stably expressing destabilized enhanced GFP (pd1EGFP) or cotransfection of plasmid with AS ODN or siRNA. Cells were plated at 40,000 cells per well in a 24-well plate and transfected 24 h later. In transient cotransfections, CHO-K1 cells were treated with 1 μg plasmid using 2 μl Lipofectamine 2000 (Invitrogen, Carlsbad, CA), as recommended by the manufacturer, together with 100 nM final concentration of ODN or siRNA, under serum-free conditions (Opti-MEM media; Invitrogen). For transfections of CHO cells stably expressing d1EGFP, the cells were similarly treated with Lipofectamine 2000 and ODN or siRNA. The transfection mixture was replaced 4 h later by fresh growth medium. Downregulation of d1EGFP in cotransfected cells (FIG. 1) and stable CHO/pd1EGFP cells (FIG. 2) was evaluated by flow cytometry at 8, 24, and 48 h following transfection. The dye, 7-aminoactinomycin D, was added to exclude nonviable cells.

Overall, the degree of antisense inhibition was greater in cotransfections compared to stable cells, whereas siRNAs were more effective in stable cells. Furthermore, for sequences complementary to the same region of the d1EGFP mRNA, AS ODNs proved more effective at the earlier time point (8 h), while siRNAs were better at subsequent time points (24 and 48 h).

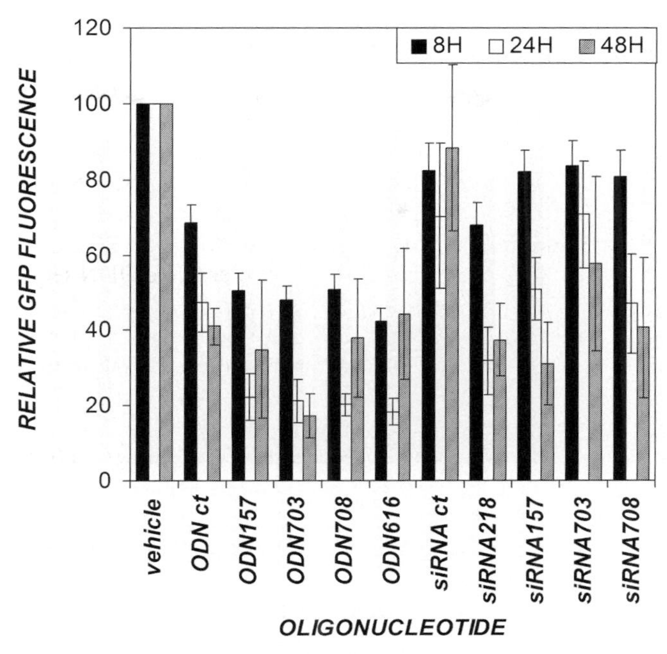

FIGURE 1. Comparison of downregulation by AS ODNs and siRNAs in transient transfections. Values shown represent the geometric mean fluorescence intensities normalized to cells transfected with pd1EGFP only (100%). Error bars represent the SEM ($n = 4$) for each group.

DISCUSSION

The thermodynamic model allowed us to select sequences that were consistently effective inhibitors for the tested panel of ODNs and siRNAs. However, thermodynamically less favorable AS ODN (616) and siRNA (218) sequences from the literature also proved to be very effective, highlighting the fact that prediction algorithms based on RNA folding calculations, with the aim of targeting regions of open secondary structure, are not perfectly accurate. Although the model selected effective AS ODNs (157, 703, and 708) using mRNA-ODN hybridization energy predictions, the effectiveness of the thermodynamically favorable matching siRNAs suggest that it can be used also to improve the selection of effective siRNA sequences. Similarly, several studies have shown that target accessibility plays an important role in the activity of AS ODNs and siRNAs.[5] However, others have suggested that while the efficacy of AS ODNs in inhibiting gene expression is governed by target accessibility and the thermodynamics of complementary bond formation, RNAi effectiveness is influenced more by the target sequence itself rather than its location (or context).[6]

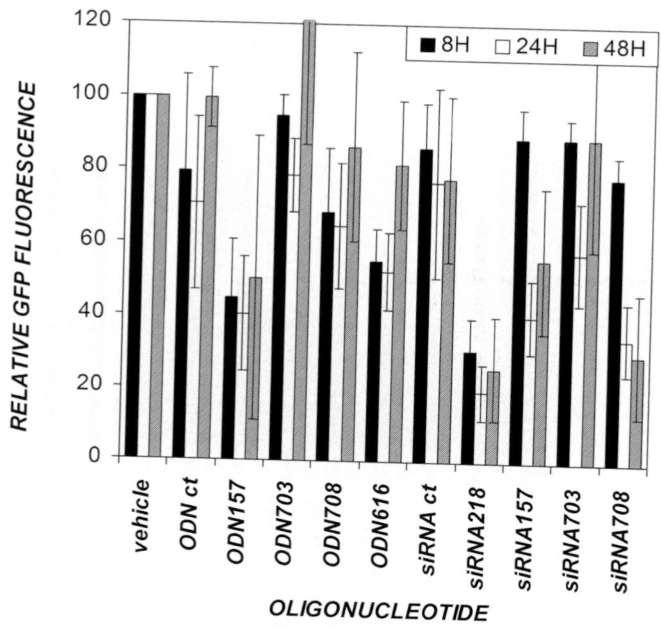

FIGURE 2. Comparison of AS ODN and siRNA mediated downregulation in stable d1EGFP-expressing cells. Fluorescence intensities are normalized to untreated control cells (100%) and are geometric mean ± SEM ($n = 4$).

The different time dependencies of antisense and RNAi activity suggest that their different modes and/or location of action could be potentially responsible. For example, siRNAs, once dissociated from the lipid carrier, are processed by the endogenous RNA-induced silencing complex, enabling access to its mRNA target. Contrary to the observation that RNAi is more effective at lower concentrations than antisense, AS ODNs were superior to siRNAs in transient transfection experiments. The mechanism for this is unclear, but it is possible that AS ODNs are inhibiting transfection of pd1EGFP as well as or instead of preventing translation of its mRNA. In our study, nonspecific inhibitory effects were detectable. Because the relationship between silencing with potency and specificity is complex and may vary with concentration, silencing dose–response experiments (at lower concentrations) should be performed in order to evaluate and improve the results. We also cannot exclude that the physicochemical properties of the lipoplexes (formed between the transfection reagent, Lipofectamine 2000, and ODNs or siRNAs) may differ between ODNs and siRNAs, and subsequently influence the cellular uptake and hence activity of the nucleic acid cargo.

REFERENCES

1. JAYARAMAN, A., S.P. WALTON, M.L. YARMUSH & C.M. ROTH. 2001. Rational selection and quantitative evaluation of antisense oligonucleotides. Biochim. Biophys. Acta. **1520:** 105–114.
2. WALTON, S.P., G.N. STEPHANOPOULOS, M.L. YARMUSH & C.M. ROTH. 1999. Prediction of antisense oligonucleotide binding affinity to a structured RNA target. Biotechnol. Bioeng. **65:** 1–9.
3. HELIN, V., M. GOTTIKH, Z. MISHAL, F. SUBRA, C. MALVY & M. LAVIGNON. 1999. Cell cycle-dependent distribution and specific inhibitory effect of vectorized antisense oligonucleotides in cell culture. Biochem. Pharmacol. **58:** 95–107.
4. BERTRAND, J.R., M. POTTIER, A. VEKRIS, P. OPOLON, A. MAKSIMENKO & C. MALVY. 2002. Comparison of antisense oligonucleotides and siRNAs in cell culture and *in vivo*. Biochem. Biophys. Res. Commun. **296:** 1000–1004.
5. KRETSCHMER-KAZEMI FAR, R. & G. SCZAKIEL. 2003. The activity of siRNA in mammalian cells is related to structural target accessibility: a comparison with antisense oligonucleotides. Nucleic Acids Res. **31:** 4417–4424.
6. YOSHINARI, K., M. MIYAGISHI & K. TAIRA. 2004. Effects on RNAi of the tight structure, sequence and position of the targeted region. Nucleic Acids Res. **32:**691–699.

Inhibition of Hepatitis C IRES-Mediated Gene Expression by Small Hairpin RNAs in Human Hepatocytes and Mice

HEINI ILVES,[a] ROGER L. KASPAR,[a] QIAN WANG,[b] ATTILA A. SEYHAN,[a] ALEXANDER V. VLASSOV,[a] CHRISTOPHER H. CONTAG,[b,c] DEVIN LEAKE,[d] AND BRIAN H. JOHNSTON[a,c]

[a]*SomaGenics, Inc., Santa Cruz, California 95060, USA*

[b]*Molecular Imaging Program at Stanford (MIPS), and Departments of Radiology, Microbiology & Immunology, Stanford University School of Medicine, Stanford, California, USA*

[c]*Department of Pediatrics, Stanford University School of Medicine, Stanford, California, USA*

[d]*Dharmacon RNA Technologies, LaFayette, Colorado 80026, USA*

ABSTRACT: The ability of small hairpin RNAs (shRNAs) to inhibit hepatitis C virus internal ribosome entry site (HCV IRES)-dependent gene expression was investigated in cultured cells and a mouse model. The results indicate that shRNAs, delivered as naked RNA or expressed from vectors, may be effective agents for the control of HCV and related viruses.

KEYWORDS: hepatitis C virus; shRNA; siRNA; RNA interference

Viral Hepatitis C is principally a disease of inflammation of the liver, and 70% of patients infected with the hepatitis C virus (HCV) develop chronic liver disease, including cirrhosis and hepatocellular carcinoma. HCV infection afflicts 3.9 million people in the United States (175 million worldwide) and is the primary indication for liver transplants in the United States. RNA interference (RNAi)-mediated gene inhibition has been shown to robustly inhibit gene expression in a number of mammalian systems.[1] Despite its high degree of sequence conservation, the HCV internal ribosome entry site (IRES) would appear *a priori* to be a poor target for RNA interference due to the high proportion of its residues involved in secondary and tertiary folding. Several groups have recently reported, however, some success targeting the HCV IRES in 293FT and Huh7 tissue culture cells (reviewed in Ref. 2).

We have investigated the ability of small hairpin RNAs (shRNAs), delivered directly or expressed from pol III promotors, as well as synthetic siRNAs to

Address for correspondence: Brian H. Johnston, SomaGenics, Inc., 2161 Delaware Ave., Santa Cruz, CA 95060. Voice: 831-426-7700; fax: 831-420-0685.
e-mail: bjohnston@somagenics.com

Ann. N.Y. Acad. Sci. 1082: 52–55 (2006).
doi: 10.1196/annals.1348.060

shRNA abbreviation	Antisense sequence (5'-3')	Target site on HCV IRES	fLuc/SEAP (at 1 nM shRNA)
HCV #1	UCAUACUAACGCCAUGGCUAGACGC	75-99	62%
HCV #2	CCGGUUCCGCAGACCACUAUGGCUC	135-159	35%
HCV #3	CCUCCCGGGGCACUCGCAAGCACCC	299-323	88%
HCV #4	UGGUGCACGGUCUACGAGACCUCCC	318-342	36%
HCV #5	UGGUGCACGGUCUACGAGACCUC	320-342	40%
HCV #6	CUCAUGGUGCACGGUCUACGAGAC	323-346	42%
HCVa-wt (#7)	UCUUUGAGGUUUAGGAUUCGUGCUC	344-368	9%
HCV #8	GGUUUUUCUUUGAGGUUUAGGAUUC	350-374	10%

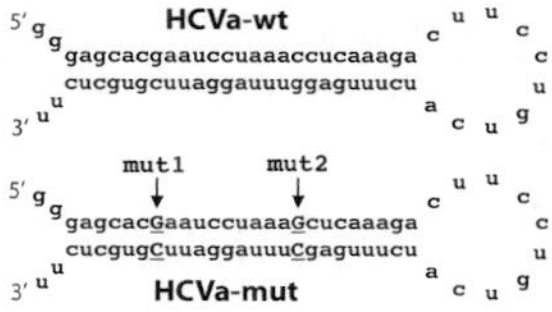

```
                HCVa-wt                  u u
5' g                                  c       c
    g gagcacgaauccuaaaccucaaaga               c
      cucgugcuuaggauuuggaguuucu               u
3' u u                                a       g
                                        c u

        mut1         mut2
5' g     ↓            ↓               u u
    g                                c       c
      gagcacGaauccuaaaGcucaaaga              c
      cucgugCuuaggauuuCgaguuucu              u
3' u u                               a       g
               HCVa-mut                c u
```

FIGURE 1. Left panel: Representative shRNA constructs assayed in the screen for potent inhibitors of HCV IRES-dependent gene expression. Antisense strands are shown with target sites on HCV IRES genotype 1a, along with the inhibitory activity of the constructs at 1 nM (for details see legend to FIG. 2). Right panel: Structure of the most potent shRNA inhibitor found, HCVa-wt. HCVa-mut1 and HCVa-mut2 contains 1 bp change, and HCVmut contains both sets of changes. All shRNAs used in the study contained the same loop sequence as well as 5′-GG and 3′-UU overhangs.

inhibit HCV IRES-dependent gene expression in cultured cells and a mouse model. We used a reporter gene plasmid in which firefly luciferase (fLuc) expression is dependent on the HCV IRES.[3] A number of shRNAs directed at various regions of the IRES were generated by *in vitro* transcription using T7 RNA polymerase (FIG. 1, left). All shRNAs had a loop derived from microRNA-23, a duplex stem of 23–25 base pairs (bp), plus 5′-GG and 3′-UU overhangs that may interact to form GU (wobble) bp (FIG. 1, right). Transfection experiments revealed the most effective shRNA, designated HCVa, which has a 25-bp stem and targets the 3′ end of the HCV IRES, near the AUG translation start site. At a concentration of 1 nM, this molecule inhibits HCV IRES-dependent luciferase expression from a cotransfected vector in 293FT cells (~90% knockdown).[3] As shown in FIGURE 2, HCVa was also highly effective in human Huh7 hepatocytes, with an IC_{50} of approximately 25 pM (FIG. 2, left). Control shRNAs containing a double mutation had little or no effect on fLuc expression, and shRNAs containing single mutations showed partial inhibition (FIG. 2, right).

shRNAs must be processed by Dicer before they can productively enter the RNA-induced silencing complex (RISC) and lead to target inhibition.[1] However, siRNAs of minimal length (~19 bp), which are most commonly used for routine gene silencing, require no such processing. To determine how shRNAs compare to minimal length siRNAs, we decided to test all possible 19-bp sequences whose target site lies within the 25-nt target site of HCVa (FIG. 3), in both siRNA and shRNA formats. Each of these RNAs showed activity at 1 nM concentration in 293FT cells, with some variability. However, the shRNAs were generally less effective than the siRNAs in this comparison, in contrast to the mouse model where shRNAs show superior efficacy, as discussed below. The poorer performance of the 19-bp shRNAs might be at least partially due to the fact that they still presumably require processing to eliminate the loop, yet may be too short to bind well to Dicer.

The inhibitors found to be active in cell culture were also evaluated in a mouse model by using *in vivo* bioluminescent imaging.[4] The fLuc expression

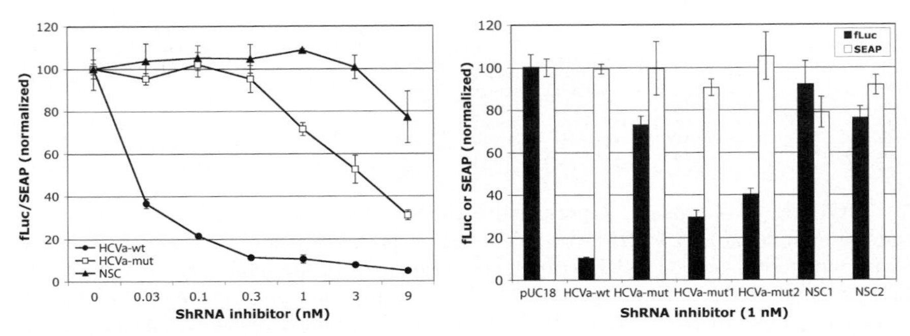

FIGURE 2. Inhibition of HCV IRES-dependent gene expression in human Huh7 hepatocytes. Huh7 cells (1.7 × 10(5) cells/well) were cotransfected (LipofectAmine 2000; Invitrogen, Carlsbad, CA) with 40 ng pHCV Dual Luc reporter construct, 50 ng pSEAP2 (as a transfection and specificity control), and the indicated amounts of shRNA inhibitors. pUC18 plasmid was added to the transfection mixture to give a total nucleic acid concentration of 800 ng per well (24-well tissue culture plates). Forty-eight hours later, supernatant was removed for SEAP analysis, and the cells were lysed and fLuc activity was measured. The data are presented as luciferase divided by SEAP activity, normalized to the pUC18 control (100%). All data were generated from individual, independent experiments performed in triplicate. NSC, nonspecific control.

plasmid, the shRNAs (or plasmid vectors expressing them), and a plasmid expressing secreted alkaline phosphatase (SEAP) were delivered to cells in the liver by hydrodynamic injection via the tail vein,[3] which leads to expression primarily in the liver. The animals were imaged following injection of luciferin at time points of 6–120 h post-injection. The results showed that the shRNA HCVa, either introduced directly or expressed by pol III from a plasmid vector, inhibited HCV IRES-dependent fLuc expression at all time points, whereas

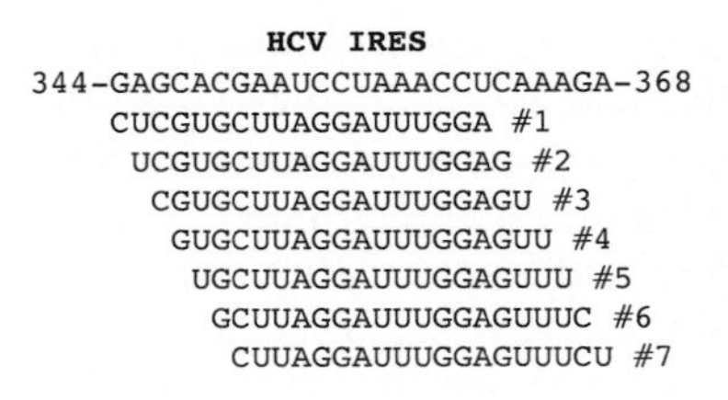

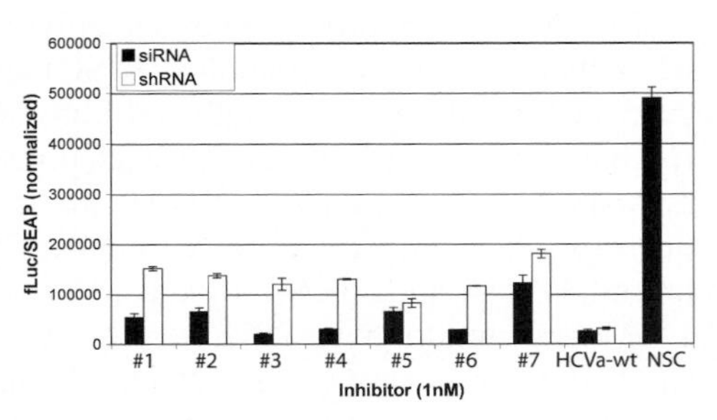

FIGURE 3. Inhibitory efficacy of all seven possible 19-bp synthetic siRNAs and *in vitro*-transcribed 19-bp shRNAs contained within the 25 nt target site for HCVa. Left panel: Target sites on the HCV IRES for the seven assayed 19-bp siRNAs and shRNAs. Right panel: 293FT cells were cotransfected with pHCV Dual Luc reporter construct, pSEAP2, and 1 nM of the indicated si/shRNA and assayed for fLuc expression (for details see legend to FIG. 2). NSC shRNA was similar to NSC siRNA.

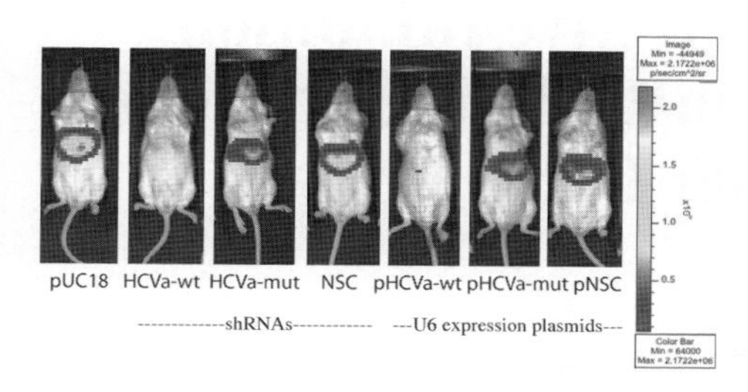

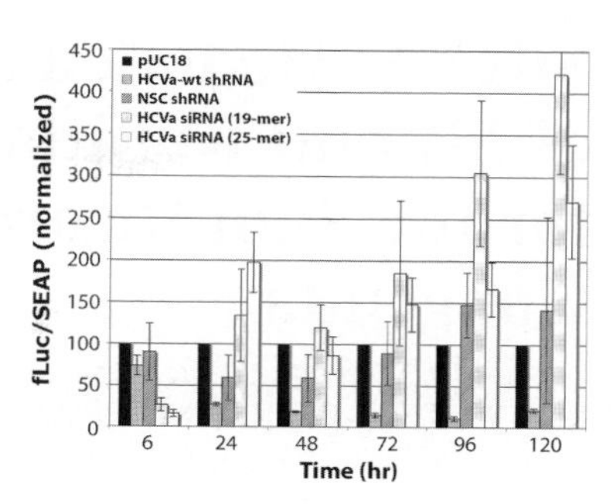

FIGURE 4. Inhibition of HCV IRES-mediated reporter gene expression in mice. pHCV Dual Luc reporter (10 μg) and pSEAP2 plasmids were coinjected at constant high pressure into the tail veins of mice with 100 μg of the indicated shRNA inhibitors or their expression plasmids. pSEAP2 provided constitutive expression of SEAP (used as a control for injection and nonspecific effects). NSC is irrelevant shRNA, and pUC18 is used as an additional control. At various time points postinjection, luciferin was administered intraperitoneally and the mice were imaged using the IVIS *in vivo* imaging system (Xenogen, Alameda, CA).[3] Representative mice from the 84 h time point are shown in the left panel; luciferase expression is shown in shaded contours. Images were quantitated using LivingImage software (Xenogen) and the time course is plotted in the right panel (4–5 mice were used per group). A total of 96 h following injection, the mice were bled and the amount of SEAP activity was determined by pNPP assay. The data are presented as luciferase divided by SEAP activity, normalized to pUC18 control mice (100%).

doubly mutated HCVa-mut or irrelevant (NSC) shRNAs had little or no effect (FIG. 4). In contrast, siRNAs showed inhibition at the earliest time point (6 h), but no inhibition was observed at later times. The lack of sustained inhibition by the siRNAs tested might be explained by the limited stability of siRNAs in blood.[1] The presence of the loop structure in the shRNAs may increase stability or may facilitate blood transport and/or cellular uptake. Taken together, these results indicate that shRNAs, delivered as naked RNA or expressed from viral or nonviral vectors, may be effective agents for the control of HCV and related viruses.

REFERENCES

1. DORSETT, Y. & T. TUSCHL. 2004. siRNAs: applications in functional genomics and potential therapeutics. Nat. Rev. Drug Discov. **3:** 318–329.
2. RADHAKRISHNAN, S.K., T.J. LAYDEN & A.L. GARTEL. 2004. RNA interference as a new strategy against viral hepatitis. Virology **323:** 173–181.
3. WANG, Q. *et al.* 2005. Small hairpin RNA efficiently inhibit hepatitis C IRES-mediated gene expression in human tissue culture cells and a mouse model. Mol. Ther. **12:** 562–568.
4. MCCAFFREY, A.P. *et al.* 2002. RNA interference in adult mice. Nature **418:** 38–39.

SiRNA-Mediated Selective Inhibition of Mutant Keratin mRNAs Responsible for the Skin Disorder Pachyonychia Congenita

ROBYN P. HICKERSON,[a] FRANCES J. D. SMITH,[b] W. H. IRWIN McLEAN,[b] MARKUS LANDTHALER,[c] RUDOLF E. LEUBE,[d] AND ROGER L. KASPAR[a]

[a]*TransDerm, Santa Cruz, California 95060, USA*

[b]*Ninewells Medical School, Dundee DD1 9SY, United Kingdom*

[c]*Rockefeller University, New York, New York 10021, USA*

[d]*Johannnes Gutenberg University, Mainz 55128, Germany*

ABSTRACT: RNA interference offers a novel approach for treating genetic disorders including the rare monogenic skin disorder pachyonychia congenita (PC). PC is caused by mutations in keratin 6a (K6a), K6b, K16, and K17 genes, including small deletions and single nucleotide changes. Transfection experiments of a fusion gene consisting of K6a and a yellow fluorescent reporter (YFP) resulted in normal keratin filament formation in transfected cells as assayed by fluorescence microscopy. Similar constructs containing a single nucleotide change (N171K) or a three-nucleotide deletion (N171del) showed keratin aggregate formation. Mutant-specific small inhibitory RNAs (siRNAs) effectively targeted these sites. These studies suggest that siRNAs can discriminate single nucleotide mutations and further suggest that "designer siRNAs" may allow effective treatment of a host of genetic disorders including PC.

KEYWORDS: siRNA; keratin disorders; pachyonychia congenita

INTRODUCTION

Pachyonychia congenita (PC) is a rare autosomal dominant negative disorder that is divided into two main subtypes, PC-1 and PC-2.[1–3] Common symptoms include hypertrophic nail dystrophy, focal palmoplantar keratoderma, blistering, oral leukokeratosis, palmoplantar hyperhidrosis, and follicular keratoses

Address for correspondence: Roger L. Kaspar, 2161 Delaware Avenue, Santa Cruz, CA 95060. Voice: 831-331-3258; fax: 831-420-0685.
e-mail: rogerkaspar@transderm.org

Ann. N.Y. Acad. Sci. 1082: 56–61 (2006).
doi: 10.1196/annals.1348.059

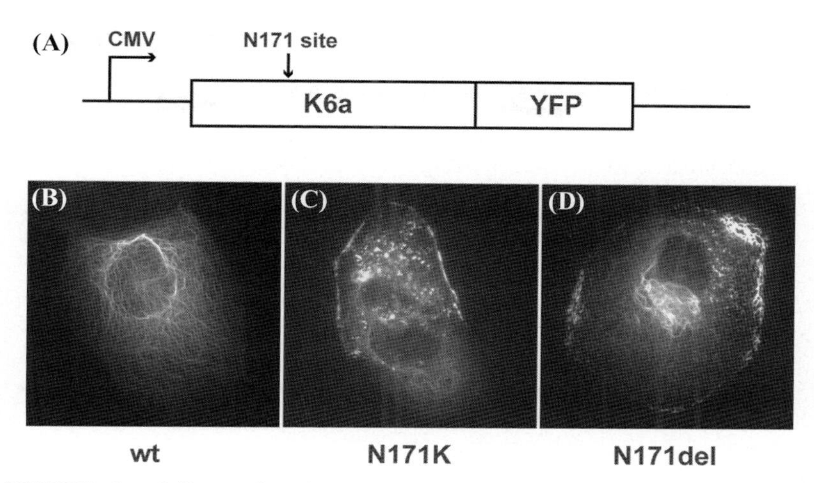

FIGURE 1. Cells transfected with plasmids expressing mutant K6a from PC patients show disrupted intermediate filaments. (**A**) Schematic of K6a (wt, N171K or N171del)/YFP fusion expression constructs. The details of the preparation of these constructs will be published elsewhere. Briefly, the human K6a insert from IMAGE clone 3639270 (MRC Geneservice, Cambridge, UK) was fused in-frame to YFP (pEYFP-N1; Clontech, Mountain View, CA) and subcloned into pcDNA5/FRT (Invitrogen, Carlsbad, CA). N171K and N171del mutations were introduced by site-directed mutagenesis and confirmed by DNA sequencing. (**B–D**) Human PLC hepatoma cells (provided by Leonard Milstone, Yale University) were transfected on a 48-well plate (~80% confluent at time of transfection) with 400 ng of pK6a(wt)/YFP (**B**), pK6a(N171K)/YFP (**C**), or pK6a(N171del)/YFP (**D**) expression construct, supplemented with pUC19 to give a final nucleic acid concentration of 800 ng/transfection, using Lipofectamine 2000 (Invitrogen) according to the manufacturer's protocol. Following transfection (24 h), the cells were trypsinized and transferred to chamber slides (Labtek II; Nunc, Rochester, NY). The cells were fixed 48 h later with 1:1 methanol/acetone, mounted and imaged by fluorescence microscopy.

on the trunk and extremities. The major complaints of PC patients are centered around pain on the pressure points of the feet following activity. This pain can be debilitating and result in patients becoming wheelchair bound.

Keratins are the type I and type II intermediate filament proteins, which form a cytoskeletal network within all epithelial cells. Mutations in these genes result in aberrant cytoskeletal networks, which present clinically as a variety of epithelial fragility phenotypes.[4] PC is known to be associated with four keratin genes. Mutations in *KRT6A* or *KRT16* lead to PC-1 and mutations in *KRT6B* or *KRT17* result in PC-2.[1–3] There are several recurrent mutations, the most common for PC-1 occurring in keratin 6a (K6a) at codon N171, which is either deleted (N171del) or a single base pair is changed resulting in an amino acid change (e.g., N171K). The recent discovery that small inhibitory RNAs (siRNAs) can effectively silence gene expression in a number of mammalian systems without inducing an immune response has resulted in intense

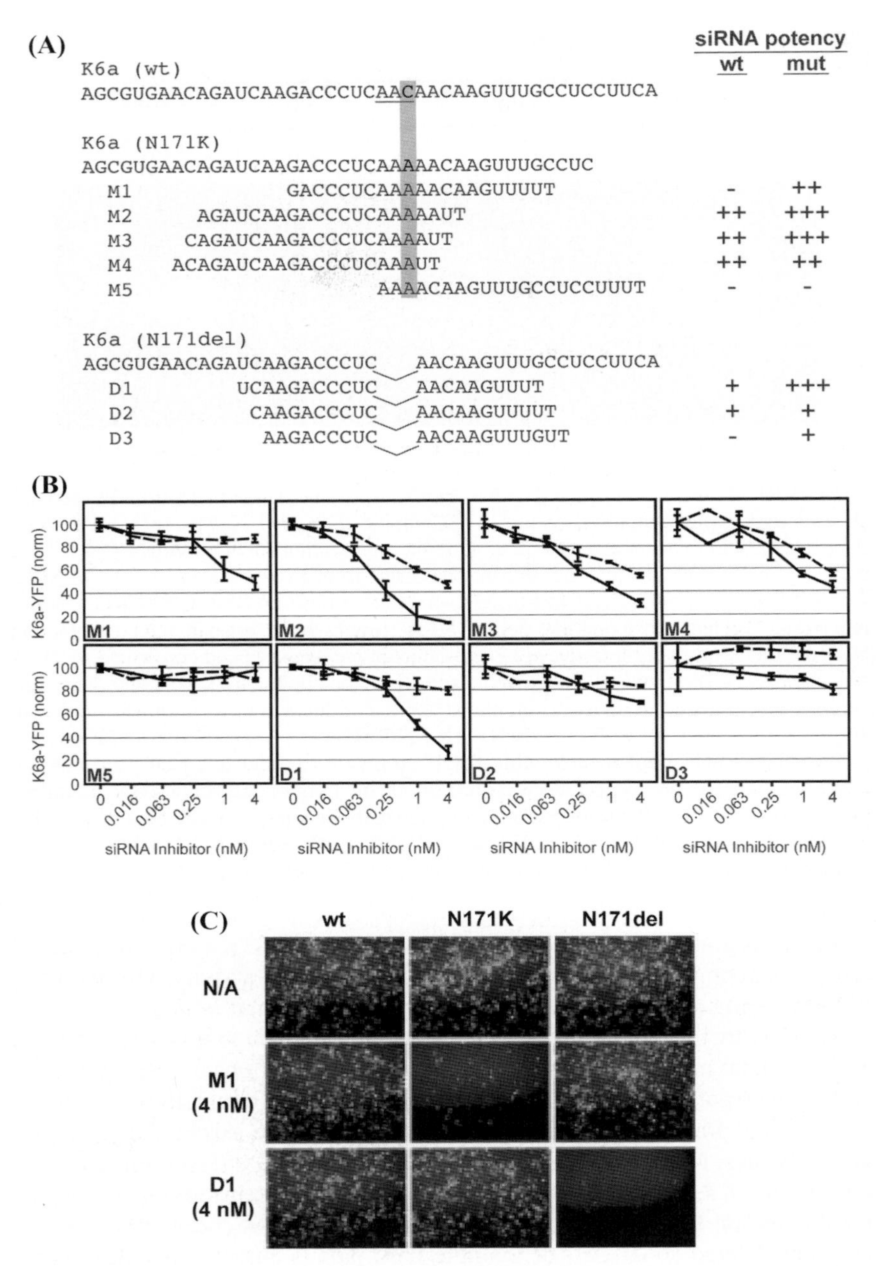

FIGURE 2. Differential inhibition of mutant versus wt K6a expression by specific siRNAs in transfected human tissue culture cells. (**A**) Sequences of siRNAs targeting K6a(N171K) (single nucleotide mutation is marked with a gray box) or K6a(N171del) (deleted nucleotides are underlined); the 3′ T is a deoxynucleotide. The wt sequence is

efforts to develop these inhibitors as disease therapeutics.[5,6] The ability to locally deliver specific, potent gene inhibitors would be a boon to patients suffering from PC and a number of other monogenic skin disorders. The focus of this article was to determine if siRNAs could be designed to inhibit expression of mutant K6a mRNA (containing either the N171K or N171del mutations) with little or no effect on wild-type (wt) K6a expression.

RESULTS AND DISCUSSION

PC-Specific Mutations Result in K6a Aggregation

In order to determine if PC-specific mutations result in disruption of keratin filament formation, human hepatoma PLC cells were transfected with wt and mutant forms of K6a fused to a reporter protein ([yellow fluorescent reporter] YFP) (FIG. 1). Introduction of K6a(wt)/YFP resulted in normal intermediate filaments in transfected PLC cells (FIG. 1 B). In contrast, similar constructs containing specific mutations in K6a derived from PC patients (N171K and N171del) resulted in keratin aggregate formation and few, if any, normal keratin filaments were observed (FIG. 1 C and D). The lack of intermediate filament integrity is thought to be responsible for the clinical symptoms of PC. Specific

←

FIGURE 2 (*Continued*). shown for comparison. Antisense sequences for M1–M5 are: 5′-AAACUUGUUUUUGAGGGUCUdT, 5′-UUUUUGAGGGUCUUGAUCUUdT, 5′-UUUUGAGGGUCUUGAUCUGUdT, 5′-UUUGAGGGUCUUGAUCUGUUdT, and 5′-AAGGAGGCAAACUUGUUUUUdT. Antisense sequences for D1–D3 are 5′-AACUUGUUGAGGGUCUUGAUdT, 5′-AAACUUGUUGAGGGUCUUGUdT, and 5′-CAAACUUGUUGAGGGUCUUUdT. SiRNA potency was quantitated from (**B**) as follows: (−) less than 10% inhibition, (+) 10–30% inhibition, (++) 30–60% inhibition, and (+++) greater than 60% inhibition. (**B**) Quantitation of fluorescence by FACS analysis. Human 293FT cells (Invitrogen) were seeded on a 48-well plate resulting in ~80% cell confluency at the time of transfection. Cells were cotransfected (in triplicate using Lipofectamine 2000) with 400 ng of the mutant (solid line) or wt (dashed line) K6a/YFP expression plasmid, 25 ng pSEAP2 plasmid (Clontech) as transfection control and the indicated amounts of synthetic siRNAs (0.016–4 nM) supplemented with pUC19 to give a final nucleic acid concentration of 800 ng. A total of 48 h following transfection, supernatant was removed for SEAP analysis.[10] The remaining cells were trypsinized and K6a/YFP expression was measured in 5000 cells by FACS (BD FACScan; BD Biosciences, San Jose, CA) using the instrument's channel FL1 (530 nm emission filter). The data were generated by gating the cells and determining the percentage of cells that dropped below the gate with or without siRNA treatment. (**C**) M1 and D1 siRNAs (4 nM) were cotransfected with K6a (wt, N171K or N171del)/YFP fusion expression constructs as described in (**B**) and visualized by fluorescence microscopy using an eGFP filter set (Chroma, Rockingham, VT). No changes were observed following siRNA treatment as observed by brightfield microscopy (data not shown).

inhibition of mutant keratin gene expression in patients may allow formation of normal keratin filaments due to the presence of one wt gene copy, thus restoring proper function.[7,8]

Differential Inhibition of Mutant versus wt K6a Expression by K6a siRNAs Targeting Mutations in Tissue Culture Cells

In order to develop potent and PC-specific inhibitors, K6a mutant-specific siRNAs designed to target PC-1 mutations N171K and N171del were tested (FIG. 2 A). SiRNAs were cotransfected into human 293FT cells with K6a(wt)/YFP or K6a(N171K)/YFP expression constructs and K6a(N171K)-specific siRNAs (M1, M2, M3, M4, and M5). M1 showed significant inhibition of mutant K6a expression (>50%) with little effect on the wt construct (FIG. 2 B). M2, M3, and M4 were effective inhibitors and showed mild discrimination between mutant and wt K6a, while M5 had no effect on either. Cotransfection of K6a(wt)/YFP or K6a(N171del)/YFP and K6a(N171del)-specific siRNAs (D1, D2, and D3) revealed high selectivity for D1 between mutant (>75% inhibition) and wt (~20% reduction) K6a/YFP expression, while D2 and D3 were ineffective against both. Little or no effect was observed on cotransfected secreted alkaline phosphatase (SEAP) expression (data not shown). These results suggest that RNA-based inhibitors can be designed and produced to be specific and highly effective against K6a mutations (including single nucleotide mutations) responsible for PC (see FIG. 2 C). Studies on epidermolysis bullosa simplex (EBS) show similar intermediate filament formation and disruption with wt and mutated K14/YFP, respectively. Furthermore, K14 mutant-specific siRNAs were able to specifically target mutant forms of the mRNA.[9] Taken together, these studies suggest that siRNAs developed against molecular targets, such as those responsible for PC, may be effective therapeutics if delivery obstacles can be overcome. The ability to quickly and cost-effectively identify siRNAs that target mutations indicate that these inhibitors may be developed as "designer" therapeutics and may help usher in the era of "individualized medicine."

REFERENCES

1. LEACHMAN, S. *et al.* 2005. Clinical and pathological features of pachyonychia congenita. J. Invest. Dermatol. Symp. Proc. **10:** 3–17.
2. SMITH, F.J.D. *et al.* 2005. Pachyonychia congenita. GeneReviews. http://www.genereviews.org
3. SMITH, F.J.D. *et al.* 2005. The genetic basis of pachyonychia congenita. J. Invest. Dermatol. Symp. Proc. **10:** 21–30.
4. OMARY, M.B., P.A. COULOMBE & W.H. MCLEAN. 2004. Intermediate filament proteins and their associated diseases. N. Engl. J. Med. **351:** 2087–2100.

5. DYKXHOORN, D.M. & J. LIEBERMAN. 2005. The silent revolution: RNA interference as basic biology, research tool, and therapeutic. Annu. Rev. Med. **56:** 401–423.
6. HANNON, G.J. & J.J. ROSSI. 2004. Unlocking the potential of the human genome with RNA interference. Nature **431:** 371–378.
7. KASPAR, R.L. 2005. Challenges in developing therapies for rare diseases including pachyonychia congenita. J. Invest. Dermatol. Symp. Proc. **10:** 62–66.
8. RUGG, E.L. *et al.* 1994. A functional "knockout" of human keratin 14. Genes Dev. **8:** 2563–2573.
9. WERNER, N.S. *et al.* 2004. Epidermolysis bullosa simplex-type mutations alter the dynamics of the keratin cytoskeleton and reveal a contribution of actin to the transport of keratin subunits. Mol. Biol. Cell. **15:** 990–1002.
10. WANG, Q. *et al.* 2005. Small hairpin RNAs efficiently inhibit hepatitis C IRES-mediated gene expression in human tissue culture cells and a mouse model. Mol. Ther. **12:** 562–568.

Multitargeted Approach Using Antisense Oligonucleotides for the Treatment of Asthma

Z. ALLAKHVERDI,[a] M. ALLAM,[a] A. GUIMOND,[b] N. FERRARI,[b] K. ZEMZOUMI,[b] R. SÉGUIN,[b] L. PAQUET,[b] AND P. M. RENZI[a,b]

[a] *CHUM Research Center, Notre-Dame Hospital, Montreal, Quebec, Canada H2L 4M1*

[b] *Topigen Pharmaceuticals Inc., Montreal, Quebec, Canada H1W 4A4*

ABSTRACT: Asthma is characterized by inflammation and hyperresponsiveness related to the accumulation of inflammatory cells, particularly eosinophils, within the airways. We tested the hypothesis that a multitargeted approach is better than a single-targeted approach in a rat model of asthma. We simultaneously delivered oligonucleotides (ODNs) targeting the chemokine receptor CCR3 and the common beta chain subunit of the receptors for IL-3, IL-5, and GM-CSF at the time of ovalbumin challenge in sensitized Brown Norway rats. Fewer eosinophils were detected in bronchoalveolar lavage (BAL) of rats treated with both ODNs as compared to each ODN alone. Moreover, airway responsiveness to LTD_4 was significantly decreased at lower doses in the 2 ODN-treated groups compared to a single ODN. As ODN therapy has raised concerns of toxicity we therefore examined ODNs prepared with modified DNA bases, specifically 2′amino, 2′deoxyadenosine (DAP) in place of adenosine. *In vivo*, administration of individual DAP-ODN was efficacious in inhibiting airway hyperresponsiveness, whereas delivery of 2 DAP-ODNs (targeting CCR3 and common beta chain) reduced the influx not only of eosinophils but also lymphocytes and macrophages in the lungs of rats as compared to the unmodified ODNs. Blocking multiple inflammatory pathways simultaneously is more effective in preventing eosinophilia and airway hyperresponsiveness than inhibiting either pathway alone. The challenges associated with the development of a product containing two oligonucleotides in humans are discussed.

KEYWORDS: asthma; common beta chain; CCR3; airway inflammation; airway hyperresponsiveness

Address for correspondence: Dr. Paolo Renzi, CHUM Research Center, Notre-Dame Hospital, 2065 Alexandre de Sève, Room Z-8905, Montreal, Quebec, Canada, H2L 2W5. Voice: 514-890-8000; ext.: 28031; fax: 514-412-7579.

e-mail: renzip@earthlink.net

Ann. N.Y. Acad. Sci. 1082: 62–73 (2006).
doi: 10.1196/annals.1348.047

INTRODUCTION

While the potential of antisense oligonucleotides (AS-ODNs) to inhibit gene expression has been identified for over 25 years,[1] challenges have arisen in the practical application of antisense approaches as therapies to treat diseases. These challenges include issues of stability of ODNs *in vivo*, uptake of the ODN by the cellular targets, and toxicity of the ODN in systemic applications.[2] With these hurdles in mind, our approach was to develop a therapy for the respiratory disease, asthma, using AS-ODN that block the expression of specific genes present in overlapping pathways important in asthma. Characteristics of asthma include airway obstruction, airway hyperresponsiveness (AHR), and accumulation and persistence of inflammatory cells in the airways, specifically the accumulation and subsequent degranulation of eosinophils.[3–5] Cytokines interleukin-3 (IL-3), IL-5, and granulocyte macrophage-colony stimulating factor (GM-CSF) are involved in this inflammatory process by mediating the survival and activation of eosinophils, as well as activating mast cells and macrophages to release proinflammatory cytokines, lipid mediators, and granules. These released products may then contribute to AHR, tissue damage, airway remodeling.[6–8] Recruitment of eosinophils and other allergic inflammatory cells to the site of inflammation is mediated by chemokines, such as eotaxin, that bind to the chemokine receptor CCR3 expressed on the cell surface of eosinophils.[9–12] CCR3 ligation has multiple effects including eosinophil recruitment, proliferation, and progenitor differentiation at the site of respiratory inflammation. Previously we demonstrated that delivery of AS-ODN targeting the common beta subunit of the IL-3/IL-5/GM-CSF receptors reduced eosinophilia and inhibited antigen-induced AHR in a rat model of asthma.[13] Recently, using the same model we have demonstrated inhibition following the delivery of AS-ODN targeting CCR3 expression.[14] With the results from these individual studies, we decided to determine whether the combined delivery of both AS-ODN together, one targeting the common beta chain of IL-3/IL-5/GM-CSF receptors and the second targeting the chemokine receptor CCR3, would function synergistically in reducing the asthma-like symptoms in the rat model. Such a combined antisense therapy required the determination of the optimal ratio of AS-ODNs, testing the possibility of duplex formation between the AS-ODN molecules in solution and assessing any increased risk of toxicity. We also pursued the advantages of chemically modifying the AS-ODN through the replacement of adenosine with 2′ amino, 2′ deoxyadenosine (DAP) to reduce potential toxic effects yet retain the efficacy of gene knockdown.

MATERIALS AND METHODS

Synthesis and Design of Rat AS-ODNs

AS-ODNs were designed in house following rules previously defined.[15] AS-ODN 143 (5′- TGG CAC TTT AGG TGG CTG – 3′) was found to be effective at inhibiting mRNA expression of the rat common beta chain receptor for IL-3/IL-5/GM-CSF both *in vitro* and *in vivo*.[13] For CCR3 antisense, AS-ODN ASA4 (5′- ACT CAT ATT CAT AGG GTG – 3′) was found to be effective at inhibiting rat CCR3 mRNA expression *in vitro* and *in vivo* studies.[14] Mismatch ODN for beta chain (MM143-ODN: 5′- GTG CCA TTT GAG TGG CTG – 3′) and A4 (MM4-ODN: 5′- CAT CAT TAT CAT GAG GTG – 3′) were also designed. All AS-ODNs were synthesized and cartridge (RP1) purified by SIGMA-Genosys (Oakville, ON, Canada). For some studies, in the synthesis of the sequences, adenosine bases were replaced by DAP, or tagged with FITC during synthesis.

Assessment of the Distribution of AS-ODNs *Within the Lungs After Intratracheal (i.t.) Administration*

Rats exposed i.t. with FITC-labeled ASA4 (200 μg) were sacrificed at different time points and immunohistochemistry was performed on cryosections from the left lungs as previously described.[14] Sections were stained with polyclonal Ab to smooth muscle actin (1:50 dilution; Biomedia Corp., Foster City, CA), counterstained with Alexa Fluoro 633-labeled goat anti-rabbit IgG2a (Invitrogen, Burlington, ON, Canada) and the nuclear stain DAPI then mounted with Prolong Antifade (Invitrogen).

OVA Sensitization and Antisense Efficacy in Airways of Challenged Rats

The study was performed at the CHUM Research Center (Montréal QC, Canada) and was approved by the resident Animal Ethic Committee. Brown Norway rats 6 to 8 weeks old (Harlan Sprague-Dawley Inc., Walkerville, MD) were sensitized by subcutaneous injection of 1 mL of saline containing 1 mg of chicken egg OVA (Sigma-Aldrich, Oakville, ON, Canada) and 3.5 mg of aluminum hydroxide gel (Laboratoire MAT, Beauport QC, Canada). Fourteen days after OVA sensitization, rats were injected i.t. with either sterile saline (50 μL) or 200 μg of AS-ODN or their corresponding MM in 50 μL sterile saline. Rats were then challenged by exposure to OVA aerosols (5% in saline) for 15 min. Fifteen hours after OVA challenge, rats were anesthetized, intubated, and baseline R_L was measured to assess the effect of AS-ODNs on AHR to leukotriene D_4 (LTD_4) (Cayman Chemical, Ann Arbor MI). Rats were then exposed to incremental doses of LTD_4 in 50 μL of saline i.t. and measurements of R_L were obtained immediately after the administration of each dose.[16,17] The concentration of LTD_4 required to double the baseline value was defined as $EC_{200}LTD_4$ and was calculated by linear interpolation between the two concentrations bounding the point at which R_L reached 200% of the baseline value.

Assessment of the effect of AS-ODNs on cellular infiltrate to the lungs was done in separate groups of rats treated as described above. Briefly, 15 h after OVA challenge, bronchoaveolar lavage (BAL) was performed with total and differential cell counts on cytospin slides stained with Hema-3 stain kit (Fisher Scientific Canada, Ottawa, ON, Canada). At least 200 cells were counted under oil immersion microscopy.

T Melt Analysis

To ensure that the human equivalents of the two oligonucleotide strands did not interact in solution, AS-ODN selected against the human sequences for beta chain (19 mer TOP004: 5′-GGG TCT GCX GCG GGX TGG T – 3′) and CCR3 (21 mer TOP005: 5′- GTT XCT XCT TCC XCC TGC CTG – 3′) (X represents where DAP replaced A) were mixed at equimolar concentrations in 1X PBS and standard UV thermo-denaturation methods were conducted using a Beckman DU640 spectrophotometer (Mississauga, ON, Canada) with a T°m accessory. Change in absorbance was detected at 260 nm at each degree from 10°C to 90°C. Melting curves were fitted using MELTWIN 3.5 software (http://www.meltwin.com/) to determine thermodynamic parameters.

Efficacy of Human AS-ODN in Cynomolgus Monkey PBMCs

PBMCs were purified from fresh blood from cynomolgus monkeys (ITR Laboratories Canada Inc., Baie d'Urfé, Quebec, Canada) by Ficoll Paque density centrifugation and transfected with different concentrations of AS-ODNs. RNA was extracted 24 to 36 h later with the RNeasy Kit (Qiagen, Freemont, CA) and using equivalent amounts of RNA (0.2 g to 2 g) RT was performed followed by PCR with an equimolar amount of cDNA. The following gene-specific primers were used: for mkCCR3, 5′ primer (5′- TGC TCT GTG AAA AAG CCG ATG- 3′) and 3′ primer (5′- ACC AAA AGT GAC AGT CCT GGC-3′); for human beta chain, 5′ primer (5′-AAG TCA GGG TTT GAG GGC TAT G-3′) and 3′ primer (5′- CAA GGG GGC AGA GAC AGG TAG-3′); and for the control GAPDH gene, 5′ primer (5′- ACC ACA GTC CAT GCC ATC AC-3′) and 3′ primer (5′- TCC ACC ACC CCT GTT GCT GTA -3′). The amount of cDNA and the number of cycles were adjusted to be in the linear range of the assay. PCR products were analyzed by ethidium bromide-stained agarose gel and quantified by densitometry using a digital camera-equipped UltraLum gel documentation system (Claremont, CA) and TotalLab software (Nonlinear USA Inc., Durham, NC), with densitometric measurements corrected for background. Level of target gene expression was normalized to levels of the control gene GAPDH. Antisense efficacy was expressed as a percentage of the reduction of gene expression relative to control (untreated) cells.

Statistical Analysis

Differences between groups were determined by analysis of variance (ANOVA). Results are expressed as mean values for the group of animals analyzed ± standard deviation (SD). Statistical significance (unpaired *t*-test) was determined as $P < 0.05$.

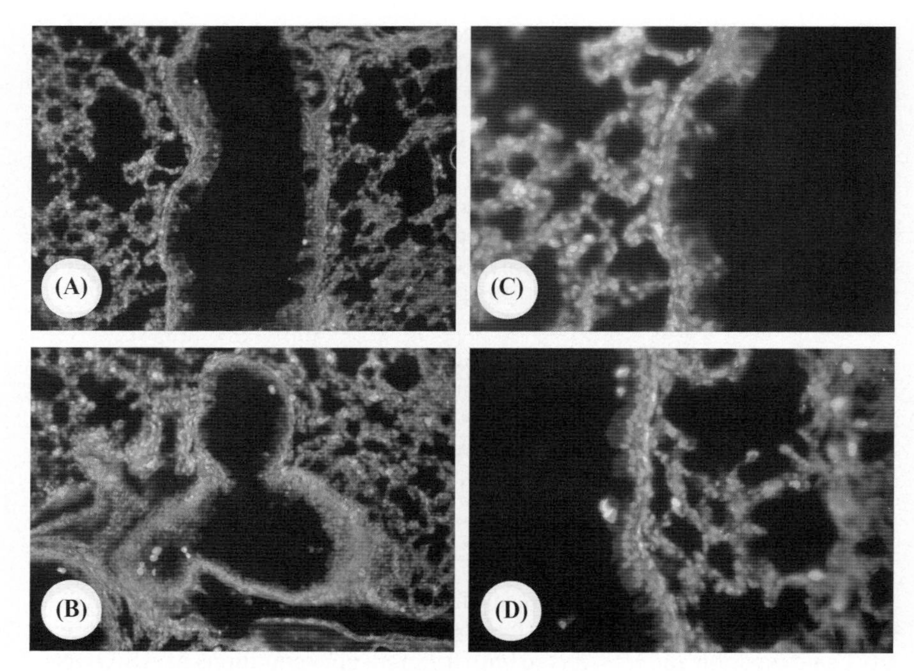

FIGURE 1. Topical delivery of tagged AS-ODN to the lungs of animals. Immunohistochemical sections of lungs of rats 24 h following i.t. exposure to FITC-labeled ASA4 (*green*). Smooth muscle actin (*red*) and nuclear label DAPI (*blue*) denote cells in the lung tissue. **A**, **B** at 200 × magnification; **C**, **D** at 400 × magnification.

RESULTS

Topical Delivery of AS-ODN to Target Tissue

As we are interested in asthma, we needed to assess that the AS-ODN can be successfully delivered to the target tissue, the lungs. Immunohistochemical sections of lungs from rats that had been exposed i.t. with fluorescently-tagged AS-ODN (200 μg) clearly demonstrated AS-ODN present in the target tissue 24 h after exposure (FIG. 1). The pattern of fluorescence was different from that of FITC administered alone to the lungs and consisted of uptake in epithelial and endothelial cells, inflammatory cells, and the airway interstitium. There was also mild uptake in smooth muscle cells.

Airway Challenged Rat Model to Determine Combination Effect of ASA4 and AS143 on AHR to LTD_4

We have previously¨ shown effective blocking of the allergen-induced eosinophilia in the lungs and reducing the AHR to LTD_4 following topical

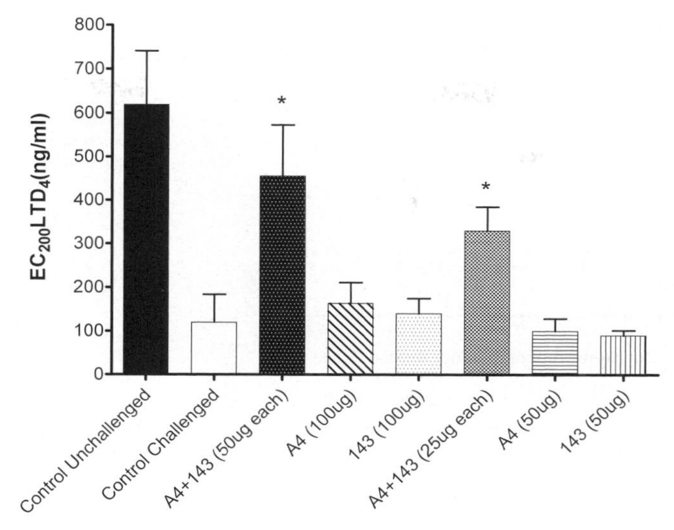

FIGURE 2. Synergistic effect of AS-ODNs targeting the common beta chain for IL-3/IL-5/GM-CSF receptor and the chemokine receptor CCR3 on antigen-induced AHR. OVA-sensitized BN rats ($n = 8$ per group) were OVA challenged and 15 h later baseline R_L was measured. Rats were exposed (i.t.) to incremental doses of LTD_4 to determine $EC_{200}LTD_4$ values. Before OVA challenge, animals received saline, AS-ODN against beta chain (AS143), AS-ODN against CCR3 (ASA4), or both AS-ODNs at indicated doses (i.t.). $^*p < 0.05$.

delivery of AS-ODN (200 μg i.t.) targeting either the common beta chain of the IL-3/IL-5/GM-CSF receptor (AS143)[13] or the chemokine receptor CCR3 (ASA4).[14] In order to evaluate whether combined delivery of both AS-ODN could act synergistically and affect AHR we determined the effect of pretreatment with both AS-ODN at doses that were shown previously to have no effect (50–100 μg).[13] As shown in FIGURE 2, OVA challenge resulted in a significant decrease in the effective dose of LTD_4 required to double the resistance baseline value ($EC_{200}LTD_4$) when compared to unchallenged rats. Single i.t. pretreatments with ASA4 or AS143 at a dose of 100 μg had no effect on the $EC_{200}LTD_4$, yielding values similar to the challenged group, however a combined pretreatment with both AS143 and ASA4 of 50 μg each together for a total dose of 100 μg significantly increased the LTD_4 effective dose required to double the resistance baseline value. This synergistic effect was observed at an even lower total combined dose of only 50 μg (25 μg each of ASA4 and AS143) although animals treated with 50 μg of either antisense alone showed no change in their $EC_{200}LTD_4$. A ratio of 1:1 of AS-ODN appeared to be effective.

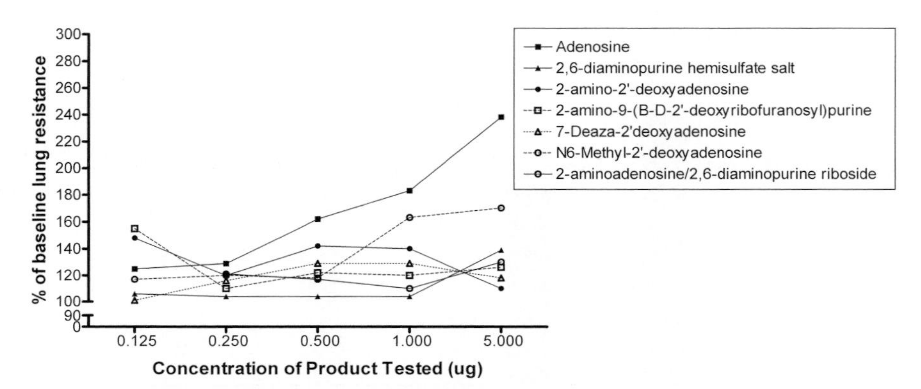

FIGURE 3. Airway responsiveness to adenosine, DAP, and DAP analogs. At 5-min intervals sensitized but unchallenged rats (≥6 per group) were administered incremental doses of adenosine (*closed squares*), DAP, or DAP analogs. The average percentage baseline lung resistance is presented.

Modification of AS-ODN with DAP

A number of chemical modifications of the nucleotides have been used in antisense development to reduce degradation by nucleases and limit toxicity. To this purpose, we compared the effects of i.t. administration of adenosine with DAP or analogs thereof on the change in baseline lung resistance in unchallenged BN rats. As shown in FIGURE 3, as the dose of adenosine increased, the average percentage baseline lung resistance in the animals also increased, whereas very little change in the average percentage baseline lung resistance was observed for increasing doses of DAP or any of its analogs. These results indicate that contrary to adenosine, DAP does not significantly affect lung resistance.

Airway Challenged Rat Model to Determine the Effects of DAP-Modified AS-ODN Combination Treatment on AHR to LTD_4 and Pulmonary Inflammation

Replacing adenosine with DAP in ASA4 and AS143 showed similar effects on AHR in the airway challenged rat model as reported in FIGURE 2. When delivered in combination, the DAP-modified ASA4 and AS143 were as effective at inhibiting AHR after allergen challenge. The increase in AHR was inhibited by 50% with the combined delivery of unmodified ASA4 and AS143 (50 μg total) and by 48% with the combination of DAP-modified ASA4 and AS143 (50 μg total). As reported with the unmodified oligonucleotides, DAP-modified AS-ODN delivered alone at a 50 μg dose had no effect on AHR in the rats (data not shown).

The effect of DAP-modified AS-ODN upon cellular influx was also studied. OVA challenge leads to an increase in the number of eosinophils in the BAL (FIG. 4) and this OVA-induced eosinophilia was inhibited in rats treated with the combination of ASA4 and AS143 when compared to rats that received no AS-ODN. Treatment with the combination of ASA4-DAP and AS143-DAP also reduced the OVA-induced pulmonary eosinophilia similar to its unmodified counterparts, these chemically modified AS-ODN also decreased the recruitment of lymphocytes and macrophages when compared to rats that did not receive any AS-ODN indicating a broader anti-inflammatory effect of the DAP-modified AS-ODN.

Testing for Duplex Formation when Combining the Selected Human AS-ODN

To ensure that the human equivalents of the two AS-ODNs did not interact in solution leading to the formation of duplexes, thermodynamic evaluations were conducted on the AS-ODN that had been selected to target the human sequences for the common beta chain and CCR3 (TOP004 and TOP005, respectively). The AS-ODNs were mixed at equimolar concentrations and standard UV thermodenaturation methods were performed to measure the absorbance at each degree change. As shown in TABLE 1, no sudden shift in the absorbance

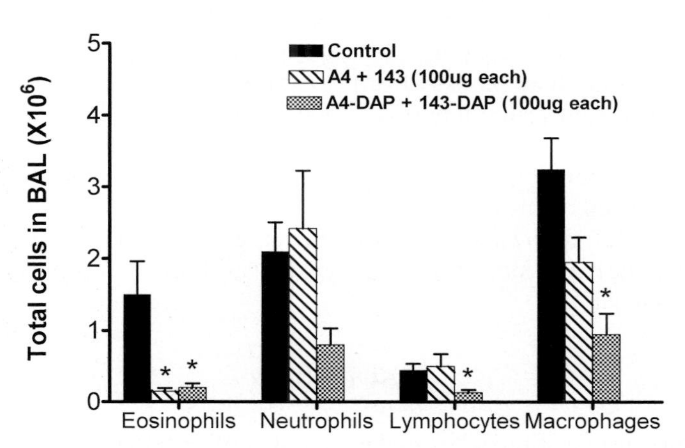

FIGURE 4. AS-ODNs containing DAP are more effective at inhibiting lung inflammation *in vivo* after challenge. Sensitized BN rats ($n = 8$ per group) were OVA challenged and 15 h later BAL was harvested and differential cell counts were assessed. Before OVA challenge, animals received (i.t.) saline, AS-ODN against beta chain composed of either only phosphorothioate (AS143) or containing DAP modifications (AS143-DAP), AS-ODN against CCR3 composed of either only phosphorothioate (ASA4), or containing DAP modifications (ASA4-DAP) or both sets of AS-ODNs at indicated doses. Data expressed as mean ± SD of specific cell types. $^*p < 0.05$ versus control.

TABLE 1. T melt analysis of human AS-ODN when combined in solution

Concentration (M)	dH (kcal/mol)	dS (cal/k-mol)	dG (kcal/mol, 37°C)	Tm (°C)
8.61e-6	–46.14	–135.02	–4.27	13.64
6.77e-6	–46.01	–134.32	–4.35	13.12
5.82e-6	–54.5	–159.85	–4.92	18.99
4.34e-6	–45.44	–130.83	–4.86	14.23
3.1e-6	–53.22	–157.27	–4.45	14.19
1.41e-6	–50.76	–145.57	–5.61	15.13

TOP004 and TOP005 mixed at equimolar concentrations in 1X PBS at different concentrations of total AS-ODN were subjected to increasing temperature and change in absorbance was detected at 260 nm at each degree from 10°C to 90°C. Melting curves were fitted using MELTWIN 3.5 software to determine thermodynamic parameters. Stability of the melting temperatures suggests that these two AS-ODNs do not form duplexes when in solution.

was observed for any of the concentrations tested, indicating that for these AS-ODNs there was no formation of duplexes at any of the temperatures or concentrations tested.

Efficacy of Combination AS-ODN in Monkey PBMCs

To verify that DAP-modified AS-ODN of the human sequence retain efficacy, TOP004 and TOP005 were tested alone and combined in a 1:1 ratio (ASM8) in cynomolgus monkey PBMCs for their ability in knocking down target gene expression. Monkey PBMCs exposed to various concentrations of AS exhibited significant inhibition in gene expression of the beta chain and CCR3 (FIG. 5). A lower dose of combined ASM8 (2.5–5 μM) was needed for maximal inhibition of gene expression (*hatched and shaded bars*, FIG. 5) when compared to cells transfected with either TOP004 (beta chain, *solid bars*) or TOP005 (CCR3, *open bars*) alone. This increased potency (synergy) of ASM8 at blocking the expression of the targets, was also observed in human cell lines (data not shown).

DISCUSSION

The two current approaches in the treatment of respiratory disease involve either the inhibition of a specific target in a single pathway or the broad and nonspecific inhibition of multiple pathways (i.e., corticosteroids). In humans, asthma is associated with the accumulation of eosinophils and other inflammatory cells.[18,19] The results presented here show effective inhibition of cellular influx and AHR following delivery of a combination of our two selected AS-ODNs: one that targets the common beta chain receptor for IL-3/IL-5/GM-CSF, and the other, the chemokine receptor CCR3, both of which

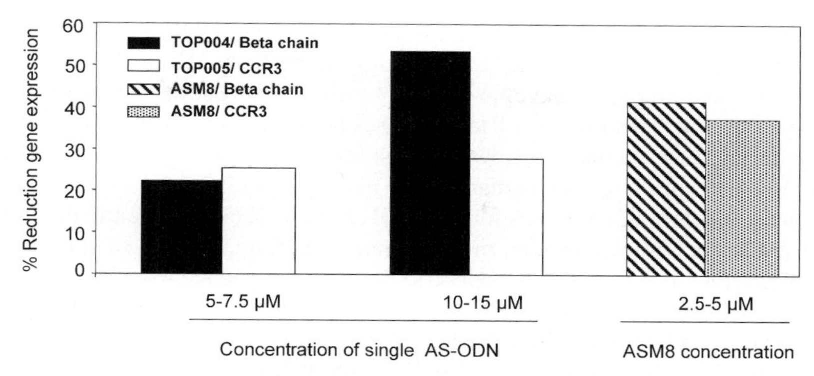

FIGURE 5. Reduced beta chain and CCR3 mRNA expression by ASM8 in cynomolgus monkey PBMCs. Purified monkey PBMCs were incubated with different concentrations of TOP004, TOP005, or ASM8 (TOP004/TOP005 in 1:1 ratio) then gene expression of either beta chain or CCR3 was assessed by RT-PCR. Results are the mean changes in gene expression obtained from experiments performed on at least five bloods of monkeys.

are mediators of allergic inflammation in asthma. The efficacy of the combined AS-ODN was observed at doses lower than either AS-ODN when delivered alone, demonstrating a synergistic effect when sets of pathways are specifically targeted.

Administration of a monoclonal antibody (mAb) targeting CCR3 was reported to partially decrease eosinophilia, but multiple doses of mAb were required for this effect and no effects on airway responsiveness were reported.[20] In another study several concurrent administrations (systemic/local) of a rat anti-mouse CCR3 mAb were required to deplete eosinophils from the lungs of OVA-sensitized/challenged mice.[21] In a clinical trial, administration of an IL-5 blocking mAb prevented blood and lung eosinophilia but had little effect on the AHR in humans with asthma.[22] As IL-3 and GM-CSF are also increased in the airways of patients with asthma, targeting IL-5 alone may be insufficient to treat asthma. We have previously reported that i.t. administration of IL-5 is only partially responsible for the physiologic changes encountered in asthma.[23] Thus limitations exist when blocking a single ligand as a therapeutic approach for asthma. In contrast, results from the present study demonstrate that a single topical administration of a low dose of combined AS-ODNs (ASA4 and AS143) was sufficient to significantly reduce both pulmonary inflammation and AHR following antigen challenge in the animal model, demonstrating the effectiveness of our approach in targeting the expression of multiple pathways rather than directly blocking a single ligand.

AS-ODNs are being developed as therapeutic agents against various diseases.[24] However, the therapeutic application of these agents may be limited as some class-related toxicity (phosphorothioate) has been described. In particular, AS-ODNs of the phosphorothiate class can bind with low affinity to a large number of proteins, which may contribute to organ distribution, accumulation, and toxicity associated with systemic delivery of AS-ODNs.[2] Thus topical delivery of AS-ODNs directly to the lungs, as performed and demonstrated in the present study, is one approach to limiting the systemic availability of AS-ODNs. Furthermore, the efficacy of these AS-ODNs delivered to the lungs may be enhanced as they are now in a position to act locally on the infiltrating immune cells and those cells marginated in lung vessels.

Many developing strategies for AS-ODN therapy focus on chemically modifying the nucleotides, sugars, or phosphate backbone to enhance stability of the AS-ODN *in vivo*, limit nonspecific binding and reduce toxicity.[2] Replacing adenosine with DAP we demonstrated similar levels of efficacy as the phosphorothiate AS-ODN on AHR in the animals as well as a broader anti-inflammatory effect suggesting that DAP modifications would enhance the overall efficacy of this therapy.

When developing a combined AS-ODN therapy using two AS-ODNs, there are additional concerns to address. The optimal ratio of each AS-ODN needs to be determined and it is important to ensure that the selected AS-ODNs are available and do not form duplexes in solution, as we demonstrated in our T°melt studies. Furthermore, in developing a human therapy it is important to be able to measure the pharmokinetics of the AS-ODNs following delivery *in vivo*. By selecting AS-ODN of different sizes (19 mer for TOP004 and 21 mer for TOP005), we were successful in determining the presence of each AS-ODN and their n-1 metabolites in the plasma of rats and monkeys even after delivery of the combined product (data not shown).

In summary, our results clearly demonstrate the possibility of combining at least two oligonucleotides into a single product and maintaining efficacy in target knockdown. Furthermore, we demonstrate that topical delivery of two AS-ODNs synergizes for potent efficacy at low doses. The modifications using DAP in this combined product not only retain the efficacy but also appear to provide a broader coverage against inflammation. We are optimistic that our AS-ODNs could be developed as a novel therapeutic for asthma.

REFERENCES

1. Zamecnik, P.C. & M.L. Stephenson. 1978. Inhibition of Rous sarcoma virus replication and cell transformation by specific oligodeoxynucleotide. Proc. Natl. Acad. Sci. USA **75:** 280–284.
2. Kurreck, J. 2003. Antisense technologies. Improvement through novel chemical modifications. Eur. J. Biochem. **270:** 1628–1644.
3. Spry, C.J. 1971. Mechanism of eosinophilia. VI. Eosinophil mobilization. Cell Tissue Kinet. **4:** 365–374.
4. De Monchy *et al.* 1995. Bronchoalveolar eosinophilia during allergen-induced late asthmatic reactions. Am. Rev. Respir. Dis. **131:** 373–376.
5. Gleich, G.J. 1990. The eosinophil and bronchial asthma: current understanding. J. Allergy Clin. Immunol. 85: 422–436.

6. ADACHIT, T. *et al.* 1995. Eosinophil viability-enhancing activity in sputum from patients with bronchial asthma: contributions of interleukin-5 and graulocyte-macrophage colony-stimulating factor. Am. J. Respir. Crit. Care Med. **151:** 618–623.
7. DRAZEN, J.M., J.P. ARM & K.F. AUSTEN. 1996. Sorting out the cytokines of asthma. J. Exp. Med. **183:** 1–5.
8. BARNES, P.J. 2003. Cytokine-directed therapies for the treatment of chronic airway diseases. Cytokine Growth Factor Rev. **14:** 511–522.
9. COMBADIERE, C., S.K. AHUJA & P.M. MURPHY. 1995. Cloning and functional expression of a human eosinophil CC chemokine receptor. J. Biol. Chem. **270:** 16491–16494.
10. GRIFFITHS-JOHNSON, D.A. *et al.* 1993. The chemokine, eotaxin, activates guinea-pig eosinophils in vitro and causes their accumulation into the lung in vivo. Biochem. Biophys. Res. Commun. **197:** 1167–1172.
11. JOSE, P.J. *et al.* 1994. Eotaxin: a potent eosinophil chemoattractant cytokine detected in a guinea pig model of allergic airways inflammation. J. Exp. Med. **179:** 881–887.
12. YANG, M. *et al.* 2003. Eotaxin-2 and IL-5 cooperate in the lung to regulate IL-13 production and airway eosinophilia and hyperreactivity. J. Allergy Clin. Immunol. **112:** 935–943.
13. ALLAKHVERDI, Z., M. ALLAM & P.M. RENZI. 2002. Inhibition of antigen-induced eosinophilia and AHR by antisense oligonucleotides directed against the common beta chain of IL-3, IL-5, GM-CSF receptors in a rat model of allergic asthma. Am. J. Respir. Crit. Care Med. **165:** 1015–1021.
14. FORTIN, M. *et al.* 2006. Effects of antisense oligodeoxynucleotides targeting CCR3 on the airway response to antigen in rats. Oligonucleotides. In press.
15. STEIN, C.A. 2001. The experimental use of antisense oligonucleotides: a guide for the perplexed. J. Clin. Invest. **108:** 641–644.
16. RENZI, P.M. *et al.* 1992. Effect of interleukin-2 on the airway response to antigen in the rat. Am. Rev. Respir. Dis. **146:** 163–169.
17. NAG, S., B. LAMKHIOUED & P.M. RENZI. 2002. Interleukin-2 induced increased airway responsiveness and lung Th2 cytokine expression occur after antigen challenge through the leukotriene pathway. Am. J. Respir. Critic. Care Med. **165:** 1540–1545.
18. WALSH, G.M. 1997. Human eosinophils: their accumulation, activation and fate. Br. J. Haematol. **97:** 701–709.
19. MOHRI, H., S. MOTOMURA & T. OKUBO. 1998. Unusual leukocytosis with eosinophilia by an allergic disease. Am. J. Hematol. **57:** 90–91.
20. GRIMALDI, J.C. *et al.* 1999. Depletion of eosinophils in mice through the use of antibodies specific for C-C chemokine receptor 3 (CCR3). J. Leukoc. Biol. **65:** 846–853.
21. JUSTICE, J.P. *et al.* 2003. Ablation of eosinophils leads to a reduction of allergen-induced pulmonary pathology. Am. J. Physiol. Lung Cell. Mol. Physiol. **284:** L169–L178.
22. LECKI, M.J. *et al.* 2000. Effects of an interleukin-5 blocking monoclonal antibody on eosinophils, airway hyperresponsiveness and the late asthmatic response. Lancet 356: 2144–2148.
23. NAG, S. *et al.* 2003. The effects of interleukin-5 on airway physiology and inflammation in rats. J. Allergy Clin. Immunol. **111:** 558–566.
24. CROOKE, S.T. 2004. Progress in antisense technology. Annu. Rev. Med. **55:** 61–95.

Therapeutic Modulation of *DMD* Splicing by Blocking Exonic Splicing Enhancer Sites with Antisense Oligonucleotides

A. AARTSMA-RUS, A.A.M. JANSON, J.A. HEEMSKERK,
C.L. DE WINTER, G.-J.B. VAN OMMEN, AND J.C.T. VAN DEUTEKOM

*Department of Human Genetics, DMD Genetic Therapy Group,
Leiden University Medical Center, 2333 AL Leiden, The Netherlands*

ABSTRACT: Antisense oligonucleotides (AONs) can be used to correct the disrupted reading frame of Duchenne muscular dystophy patients (DMD). We have a collection of 121 AONs, of which 79 are effective in inducing the specific skipping of 38 out of the 79 different DMD exons. All AONs are located within exons and were hypothesized to act by steric hindrance of serine-arginine rich (SR) protein binding to exonic splicing enhancer (ESE) sites. Indeed, retrospective *in silico* analysis of effective versus ineffective AONs revealed that the efficacy of AONs is correlated to the presence of putative ESE sites (as predicted by the ESEfinder and RESCUE-ESE software). ESE predicting software programs are thus valuable tools for the optimization of exon-internal antisense target sequences.

KEYWORDS: antisense oligonucleotides; AONs; Duchenne muscular dystophy; exonic splicing enhancer; ESE

The dystrophin protein anchors cytoskeletal actin to the sarcolemma of muscle fibers.[1,2] Frame disrupting mutations in the DMD gene result in nonfunctional dystrophins and are thus causing progressive muscle degeneration in Duchenne muscular dystrophy (DMD).[3,4] In contrast, internal deletions that keep the reading frame intact allow the generation of partially functional dystrophins, as found in the less severely affected Becker muscular dystrophy (BMD) patients.

We are exploring antisense oligonucleotides (AONs) as small molecule drugs to induce the skipping of specific exons in mutated transcripts.[5,6] This allows restoration of the reading frame and the generation of BMD-like dystrophins. We have confirmed the therapeutic applicability of this approach in cultures derived from 11 DMD patients, carrying different deletions, nonsense

Address for correspondence: J.C.T. Van Deutekom, Wassenaarseweg 72, 2333 AL Leiden, The Netherlands. Voice: +31-71-5276080; fax: +31-71-5276075.
e-mail: Deutekom@lumc.nl

**Ann. N.Y. Acad. Sci. 1082: 74–76 (2006). © 2006 New York Academy of Sciences.
doi: 10.1196/annals.1348.058**

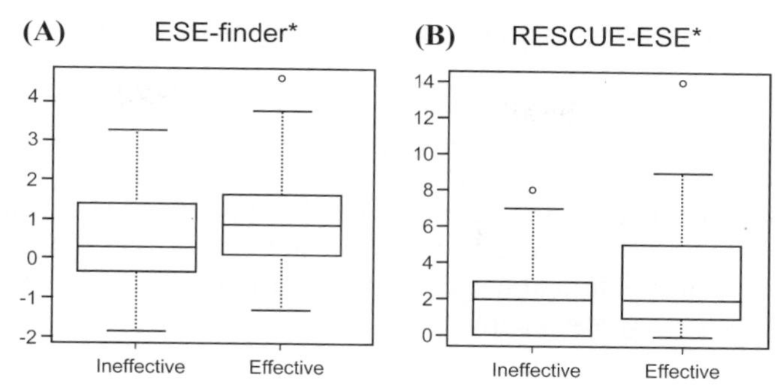

FIGURE 1. Boxplots of the respective values predicted by ESEfinder and RESCUE-ESE for effective and ineffective AONs. (**A**) ESEfinder values (adapted from Ref. 13). For each AON the most likely putative SR binding motive was used (i.e., the highest value over threshold). Effective AONs have significantly higher predicted values than ineffective AONs (*P* value 0.037, as calculated with a one-sided Wilcoxon signed rank sum test). (**B**) RESCUE-ESE values. Effective AONs contain significantly more ESE hexamers than ineffective AONs (*P* value 0.036, as calculated with a one-sided Wilcoxon signed rank sum test).

mutations or duplications[7–9] (unpublished results). First-in-man studies where AONs will be applied to Duchenne patients have now become within reach.

Exon-internal AONs are hypothesized to act by steric hindrance of SR proteins. This is a subgroup of splicing factors that recognize exonic splicing motives (so-called exonic splicing enhancer [ESE] sites).[10] SR proteins mediate splicing by recruiting other splicing factors, such as U1 snRNP and U2AF to the splice sites. Recently two computer programs to predict putative ESE sites have become available. Sequence motives recognized by the four most abundant SR proteins (SF2/ASF, SC35, SRp40 and SRp55) are implemented in the ESEfinder software.[11] The RESCUE-ESE software on the other hand contains 238 hexamers that are significantly more often found in exonic than in intronic sequences and in addition enriched in exons with weak consensus splice sites.[12]

We have designed a total of 121 exon-internal AONs targeting 38 different DMD exons [5,13] (unpublished results). The design of these AONs was primarily based on open regions in the secondary pre-mRNA structure (as predicted by the m-fold software). Accordingly, 65% of these AONs were found to be effective, albeit at variable levels. To assess whether or not there is a correlation between AON efficacy and the presence of putative ESEs in the target sequence, we performed retrospective *in silico* analysis of the entire AON collection using ESEfinder and RESCUE-ESE software.[13] We indeed observed significantly higher predicted SR protein binding site values for effective AONs when compared to ineffective AONs using ESEfinder (FIG. 1 A).[13] In addition, the number of ESE hexamers as predicted by RESCUE-ESE was significantly increased for effective AONs when compared to ineffective AONs (FIG. 1 B).

These results indicate that optimal antisense target sequences in open parts of the pre-mRNA secondary structures may be identified with ESE predicting software. Furthermore, even though the mechanism of exon-internal AON-mediated splicing modulation is not yet entirely elucidated, these data suggest that the blocking of SR binding sites plays a significant role. However, other factors, such as the timing and sequence of pre-mRNA splicing and the strength of the splice sites may also be involved.

REFERENCES

1. KOENIG, M. *et al.* 1988. The complete sequence of dystrophin predicts a rod-shaped cytoskeletal protein. Cell **53:** 219–226.
2. YOSHIDA, M. *et al.* 1990. Glycoprotein complex anchoring dystrophin to sarcolemma. J Biochem. (Tokyo) **108:** 748–752.
3. HOFFMAN, E.P. *et al.* 1987. Dystrophin: the protein product of the Duchenne muscular dystrophy locus. Cell **51:** 919–928.
4. MONACO, A.P. *et al.* 1988. An explanation for the phenotypic differences between patients bearing partial deletions of the DMD locus. Genomics **2:** 90–95.
5. AARTSMA-RUS, A. *et al.* 2002. Targeted exon skipping as a potential gene correction therapy for Duchenne muscular dystrophy. Neuromuscul. Disord. **12:** S71–S77.
6. VAN DEUTEKOM, J.C. *et al.* 2003. Advances in Duchenne muscular dystrophy gene therapy. Nat. Rev. Genet. **4:** 774–783.
7. AARTSMA-RUS, A. *et al.* 2003. Therapeutic antisense-induced exon skipping in cultured muscle cells from six different DMD patients. Hum. Mol. Genet. **12:** 907–914.
8. AARTSMA-RUS, A. *et al.* 2004. Antisense-induced multiexon skipping for duchenne muscular dystrophy makes more sense. Am. J. Hum. Genet. **74:** 83–92.
9. VAN DEUTEKOM, J.C. *et al.* Antisense-induced exon skipping restores dystrophin expression in DMD patient derived muscle cells. Hum. Mol. Genet. **10:** 1547–1554.
10. CARTEGNI, L. *et al.* 2002. Listening to silence and understanding nonsense: exonic mutations that affect splicing. Nat. Rev. Genet. **3:** 285–298.
11. CARTEGNI, L. *et al.* 2003. ESEfinder: A web resource to identify exonic splicing enhancers. Nucleic Acids Res. **31:** 3568–3571.
12. FAIRBROTHER, W.G. *et al.* 2002. Predictive identification of exonic splicing enhancers in human genes. Science **297:** 1007–1013.
13. AARTSMA-RUS, A. *et al.* 2005. Functional analysis of 114 exon-internal AONs for targeted DMD exon skipping: indication for steric hindrance of SR protein binding sites. Oligonucleotides **15:** 284–297.

Neuromuscular Therapeutics by RNA-Targeted Suppression of ACHE Gene Expression

AMIR DORI[a] AND HERMONA SOREQ[b]

[a]*Department of Neurology, Soroka University Medical Center, Ben-Gurion University of the Negev, Beer-Sheva, Israel 84105*

[b]*The Life Sciences Institute, The Hebrew University of Jerusalem, Jerusalem, Israel 91904*

ABSTRACT: RNA-targeted therapeutics offers inherent advantages over small molecule drugs wherever one out of several splice variant enzymes should be inhibited. Here, we report the use of Monarsen, a 20-mer acetylcholinesterase-targeted antisense agent with three 3′-2′o-methyl-protected nucleotides, for selectively attenuating the stress-induced accumulation of the normally rare, soluble "readthrough" acetylcholinesterase variant AChE-R. Acetylcholine hydrolysis by AChE-R may cause muscle fatigue and moreover, limit the cholinergic anti-inflammatory blockade, yielding inflammation-associated pathology. Specific AChE-R targeting by Monarsen was achieved in cultured cells, experimental animals, and patient volunteers. In rats with experimental autoimmune myasthenia gravis, oral delivery of Monarsen improved muscle action potential in a lower dose regimen (nanomolar versus micromolar), rapid and prolonged manner (up to 72 h versus 2–4 h) as compared with the currently used small molecule anticholinesterases. In central nervous system neurons of both rats and cynomolgus monkeys, systematic Monarsen treatment further suppressed the levels of the proinflammatory cytokines interleukin-1 (IL-1) and IL-6. Toxicology testing and ongoing clinical trials support the notion that Monarsen treatment would offer considerable advantages over conventional cholinesterase inhibitors with respect to dosing, specificity, side effects profile, and duration of efficacy, while raising some open questions regarding its detailed mechanism of action.

KEYWORDS: anticholinergic treatment; antisense treatment; myasthenia gravis; neuromuscular junction; readthrough acetylcholinesterase

INTRODUCTION

Cholinergic neurotransmission is a key function in both cholinergic brain synapses and at neuromuscular junctions (NMJ). Presynaptic release of

Address for correspondence: Hermona Soreq, The Life Sciences Institute, The Hebrew University of Jerusalem, Jerusalem, Israel 91904. Voice: +972-2-658-5109; fax: +972-2-652-0258.
e-mail: soreq@cc.huji.ac.il

Ann. N.Y. Acad. Sci. 1082: 77–90 (2006). © 2006 New York Academy of Sciences.
doi: 10.1196/annals.1348.004

acetylcholine (ACh) into the synaptic cleft and its binding and activation of acetylcholine receptors (AChR) located on the pre- and postsynaptic membranes, induce an action potential characteristic of cholinergic transmission. At the postsynaptic site, AChR transduce the signals through ion channel activation and secondary messenger pathways. Cholinergic neurotransmission is terminated with the catalytic hydrolysis of ACh by the carboxylesterase enzyme acetylcholinesterase (AChE).[1] Nevertheless, regulation of the balance between ACh and AChE levels and their net effect on postsynaptic stimulation is simultaneously governed by multiple mechanisms.

CHOLINERGIC HYPEREXCITATION AND INDUCTION OF THE "READTHROUGH" AChE-R VARIANT

Overt cholinergic stimulation occurs following acute psychological stress, similar to acute intoxication with AChE inhibitors, both expressed as neuronal hyperexcitability.[2,3] This intense cholinergic overdrive in both brain and NMJ synapses facilitates emergent regulatory actions. Within a few minutes, it induces a dramatic increase in c-Fos mRNA,[4,5] which leads to the activation of c-Fos binding sites on promotors of the ACh-synthesizing enzyme choline acetyltransferase (ChAT),[6] the vesicular ACh transporter (VAChT),[7] as well as the *ACHE* gene.[8] A decrease in ChAT and VAChT then suppresses ACh synthesis and packaging, while an increase in AChE mRNA enhances ACh hydrolysis, resulting with an effective restraint of the hypercholinergic response.

The increased AChE activity observed in both brain and NMJ following AChE inhibition is due to induction of a soluble AChE variant[4] (FIG. 1 A). This soluble variant is produced by alternative splicing at the 3′-end of the AChE pre-mRNA, in which pseudo-intron 4 is retained in the mature transcript, and hence termed *readthrough* AChE (AChE-R).[9] AChE-R shares similar ACh hydrolyzing efficacy with the major "synaptic" AChE-S variant due to a common core domain, but carries a distinct C-terminal extension devoid of cysteine,[10] rendering it a hydrophilic soluble monomer.[11]

In contrast to AChE-R induction, the multimeric membrane-bound AChE-S variant maintains unmodified levels, reflecting alternative splicing of AChE pre-mRNA. This profound feedback response is controlled by the splice factor SC35[12] and serves well in the short term to quickly reduce the cholinergic hyperactivity following an acute insult.[13]

AChE-R mRNA levels remain elevated for weeks following a few days of exposure to very low doses of the AChE-inhibitor diisopropylfluorophosphonate (DFP).[5,14] In the hippocampus, additional exposure to AChE inhibitors could not be rebalanced by subsequent cholinergic stimulation due to limited ability for additional AChE-R induction.[14] Therefore, repeated cholinergic stimulation by AChE inhibition induces long-term hypersensitization with potentially damaging implications. Supporting this notion,

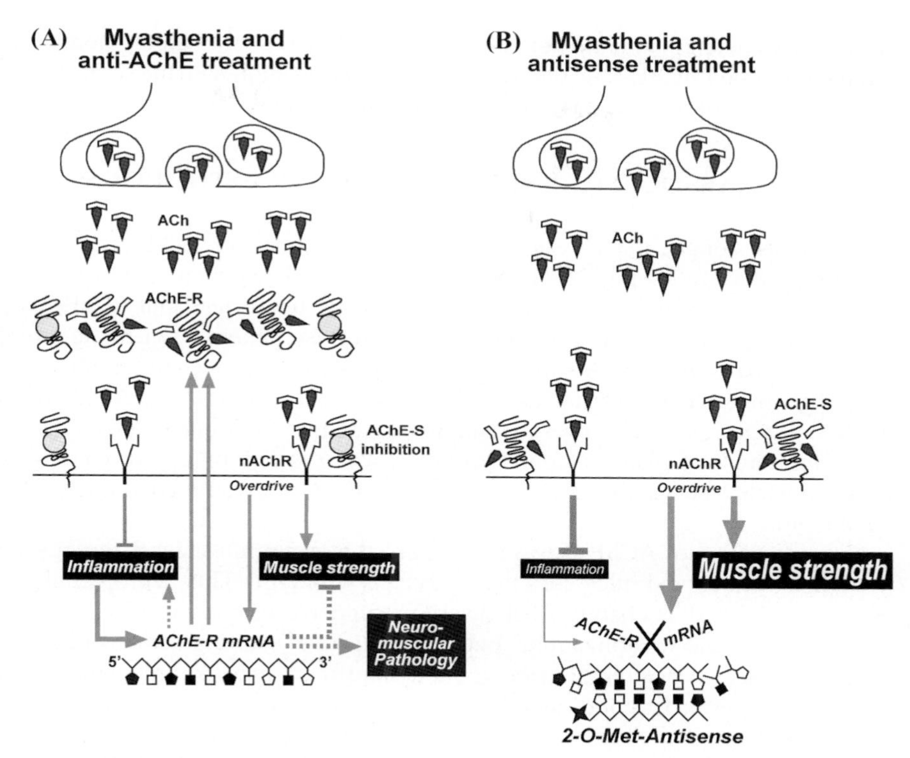

FIGURE 1. The reciprocal influence of AChE-R on AChR activation and its interferences by antisense treatment. Schematically represented is the presynaptic release of 12 ACh molecules that attempt to stimulate two nAChRs. When AChR is overtly stimulated, as occurs due to AChE inhibition (**A**), induction of AChE-R expression results. Soluble AChE-R saturates the synaptic cleft and surrounding region, efficiently hydrolyzing ACh, with only six molecules remaining for AChR activation. This results with muscle weakening (*red dashed line*) due to reduced cholinergic transmission to muscles, and induction of neuromuscular pathology; and additional loss of the inhibitory effect of ACh on cytokine release, that is, a proinflammatory response (*green dashed line*). Antisense treatment (**B**) induces degradation of AChE-R mRNA, efficiently attenuating production of AChE-R. Uninhibited AChE-S hydrolyses ACh but without saturation of the synaptic cleft with AChE-R, more ACh remains and AChR is activated by eight molecules of ACh. The muscle weakening and proinflammatory effect of AChE-R is withdrawn by increased levels of ACh. AChE-R neuromuscular pathology is ameliorated. ACh, Acetylcholine; nAChR, nicotinic Acetylcholine receptor; AChE-R, Acetylcholinesterase "readthrough" variant; AChE-S, Acetylcholinesterase "synaptic" variant. Arrows indicate direction of effect, with thickness indicating intensity.

transgenic mice overexpressing AChE-S in the central nervous system present delayed cognitive and neuroanatomical pathologies.[15,16] In humans, traumatic stress with severe stress response is often followed by long-term pathological changes,[17,18] and in extreme cases, such changes are clinically referred to as the anxiety-associated posttraumatic stress disorder (PTSD).[19] In healthy human volunteers, blood AChE-R levels are increased in association with the subjects' state anxiety,[20] suggesting physiological relevance to these increases.

AChE-R-ASSOCIATED NEUROMUSCULAR PATHOLOGY

The mouse tongue muscle shows a significant increase in AChE catalytic activity 2 weeks following DFP exposure for 4 days.[5] The punctuated expression pattern of NMJ-related AChE-S mRNA remains unchanged, but AChE-R mRNA is significantly induced and exhibits a diffuse pattern, suggesting an extrajunctional synthesis. A similar expression pattern was detected in transgenic TgS mice overexpressing human AChE-S, which also show host AChE-R mRNA overexpression.[5] Corresponding to this, activity staining localized the signal to motor end plates in control animals, whereas TgS mice showed dispersed muscle fibers staining of the soluble AChE-R, not restricted to the end plate region.

Overexpression of AChE-R was accompanied by neuromuscular pathology in both TgS mice and mice chronically treated with DFP. This neuropathology was exhibited as chaotic fiber disorder compared to control mice, with severe atrophy and vacuolization.[5] Furthermore, silver staining demonstrated unaffected large nerve bundle fibers, but a significant increase in small unbundled neurites in TgS mice and following chronic DFP treatment compared to controls. Corresponding to the already known neurite-promoting activity of AChE,[21] this indicates axon branching under the influence of AChE-R overexpression, and suggests that a denervation–reinnervation process occurs in these affected muscles. Furthermore, nontransgenic mice treated with DFP and TgS mice exhibited a significant increase in the number of end plates in their diaphragm muscle compared to controls.[5] These were observed to be smaller in size, collectively reinforcing the assumption that reinnervation occurred.

Examination *ex vivo* of intact TgS muscles demonstrated rapid fatigue following initial stimulation, increased decrement, and delayed recovery compared to strain-matched controls when stimulated directly via the phrenic nerve.[15] This was attributed to both neuronal and muscle impairments.

ANTISENSE SUPPRESSION OF AChE-R EXPRESSION

The muscle-weakening effects of AChE-R overexpression could be alleviated by small molecule inhibitors, but the effect was short lasting. This could be due to the failure of small molecule inhibitors to distinguish between the different AChE variants, which would induce general increase in ACh and subsequent feedback overproduction of yet more AChE-R. Supporting this notion, the effect of such small molecule inhibitors to improve muscle functioning in human patients is also short lasting, to the extent that patients typically require oral drug administration every 4–6 h, and up to every 3 h in more refractory cases.[22,23] Antisense oligonucleotide suppression of nascent AChE-R mRNA transcripts emerged as a promising alternative. Designated AS3[16,24–26] or EN101[27,28] and recently termed *Monarsen* for primate AChE,[29] the AChE mRNA-targeted agent is a 20 residues antisense molecule, 2′-oxymethylated at its three 3′-terminal positions. It is targeted at exon 2 of AChE mRNA, which is common to the different AChE mRNA splice variants.[9] Nevertheless, the nascent, relatively unstable AChE-R transcripts demonstrated particular sensitivity to the antisense treatment compared to AChE-S mRNA in cultured cells,[14] live mice,[16,30,31] myasthenic rats,[27] human-originated cell lines,[32,33] and orally treated primates.[29] The preferred destruction of AChE-R mRNA could be due to its being overproduced under stress or disease, not yet protected by polyribosomes. Alternatively, or additionally, it could be due to its being longer, and less G,C-rich than AChE-S mRNA,[34] rendering it relatively unstable.[35]

AChE catalytic activity was increased in muscle 2 weeks following repeated DFP exposure for 4 consecutive days. This increase reflected overexpressed AChE-R and could be significantly reduced by antisense treatment, which brought down the catalytic activity to the range of control animals.[5] Furthermore, antisense treatment prevented the increase in NMJ end plates density, further indicating that these plastic changes were attributed to AChE-R. The AChE-R-related muscle weakness resembles the reported weakness syndrome that follows continuous or repeated exposure to agricultural anti-AChE pesticides, such as the organophosphate paraoxon[36,37] or warfare agents, such as those associated with the Gulf War syndrome.[38]

AChE-R OVEREXPRESSION IN MYASTHENIA GRAVIS

A disease where the currently accepted treatment involves administration of an anti-AChE small molecule is myasthenia gravis. In this disease, autoantibodies against nicotinic AChR induce NMJ malfunction.[39] This reduces responsiveness to ACh, resulting with limited muscle contraction following stimulation of the motor nerve, and ensuing weakness and fatigue. Similar symptoms may be detected in rats with experimental autoimmune myasthenia gravis (EAMG), induced by injection of purified AChR that induces production of autoantibodies. In EAMG rats as well as in myasthenic patients, in spite of low cholinergic output of the NMJ, the cholinergic imbalance with dominance of ACh over AChR results with induction of AChE-R and its accumulation in serum.[27] This may be due to intense activation of AChR, as occurs in the normal NMJ under anti-AChE treatment. In EAMG rats, skeletal muscles were found to be depleted of nicotinic AChR, and the synaptic bound AChE-S remained unchanged both in its expression level and in its clustered localization. In contrast, these muscles were saturated with induced AChE-R that showed a dispersed cytoplasmic localization, as appropriate for its secretory–soluble characteristics. This indicates that while AChE-S hydrolyses ACh in the synaptic cleft, AChE-R functions as a scavenger, eliminating ACh around the end plate, and furthermore—in plasma, efficiently limiting AChR "over" activation.

ANTISENSE TREATMENT OF MYASTHENIA GRAVIS

The primary rational for applying antisense treatment against AChE-R in myasthenia gravis was the attempt to reduce the junctional and perijunctional excessive ACh hydrolysis by AChE-R. Antisense treatment significantly reduced AChE-R mRNA and protein in EAMG muscle and plasma, while AChE-S remained unaffected. This treatment did not affect AChR levels, attesting to its specificity. The fatigue, which is a hallmark of myasthenic patients is similarly measurable in EAMG rats by electromyography, demonstrating a decrement, with a decrease in compound muscle action potential (CMAP) during repeated stimulation at 3 Hz. Antisense treatment did not affect CMAP in control animals, but in EAMG rats, it normalized the CMAP within 1 h and in nanomolar concentrations. This was accompanied by increased mobility, correction of posture, enhancement of grip, and reduction of tremulousness during ambulation, which are all features that indicate enhanced muscle force. Corresponding to this, treated rats demonstrated a significantly improved performance on a treadmill, while healthy animals were unaffected by the treatment.[27]

The extent and the duration of CMAP correction by antisense treatment were both dose-dependent. For example, 500 μg/kg increased a CMAP up to 125% of base line for over 72 h, whereas a 50 μg/kg dose was less effective and only lasted 24 h. In both cases, the effect was considerably longer than that offered by the currently used small molecule drug, compatible with the advantage of selectively suppressing AChE-R levels. EN102, which is a similar 3′-protected antisense oligonucleotide, targeting another sequence in AChE mRNA (AS1 in Ref. 24), produced a similar improvement of CMAP decrements in EAMG rats, further indicating that AChE-R overexpression significantly contributed to these decrements. Comparable amounts of an inverted sequence—invEN102, with no target in the database, did not improve muscle function, attesting to the sequence specificity of the antisense (EN101) effect. Dose–response curves revealed that up to 5 h after an injection, EN101 produced a saturable response with an IC_{50} of <10 μg/kg. This effect appeared to be superimposed on a much longer lasting and less concentration-dependent effect that showed no saturation in the range studied, possibly reflecting altered muscle and/or NMJ properties under the stable CMAP retrieval afforded by the antisense treatment. As 2′-oxymethyl-protected antisense oligonucleotide agents are efficient under oral administration,[40] EN101 and EN102 were further administered orally. A dose of 50 μg/kg once a day via an intubation feeding needle was as effective as

25 μg/kg EN101 administered intravenously, though EN102 effects appeared somewhat delayed compared with EN101.

NONCHOLINERGIC EFFECTS OF AChE-R ON THE MYASTHENIC NMJ

Conventional anti-AChE treatment administered to EAMG rats corrected CMAP decrements more rapidly, though to a shorter interval compared to the antisense treatment. Nevertheless, antisense-treated EAMG rats exhibited a significantly improved survival throughout a month of treatment compared to conventional myasthenia treatment with pyridostigmine, with better treadmill performance and weight gain. This may suggest that the antisense treatment that reduces AChE-R modifies the NMJ pathophysiology and possibly the course of the disease. Indeed, the conventional anti-AChE treatment is known to promote structural abnormalities of the postsynaptic NMJ membrane, terminal nerve branching, and myopathologies in rats and mice.[5,41,42] Therefore, it is conceivable that the treatment itself may induce progressive NMJ pathology. At least some of these changes may be attributed to AChE-R induction[5] by the nondiscriminative small molecule inhibitors. In contrast, antisense suppression of AChE-R likely avoids this iatrogenic deterioration and the progressive course of the disease by selectively destroying the target chains and sparing the neuromuscular enzyme (FIG. 1 B).

AN ADDED VALUE: MODIFIED CYTOKINE EXPRESSION BY ANTISENSE SUPPRESSION OF AChE-R

Recently, the "cholinergic anti-inflammatory pathway" was identified, whereby the cholinergic nervous system modulates the inflammatory response through ACh.[43,44] ACh activates homomeric α7 nicotinic AChR on tissue-residing macrophages and thereby attenuates the secretion of proinflammatory cytokines.[45] Regulation of this cholinergic immunemodulation is obviously essential for appropriate homeostasis; and induced AChE-R that efficiently scavenges ACh in the brain, within and around neuromuscular junctions and in plasma was correspondingly predicted to have a significant role, leading to increased production of proinflammatory cytokines. Additionally, increased inflammatory cytokine levels, for example, interleukin-1 (IL-1),[46] may exert secondary influence by increasing AChE gene expression throughout the nervous system, forming a closed gap loop and prolonging this proinflammatory effect.

In cynomolgus monkeys, stress induced by oral gavage administration of a low dose of antisense against AChE mRNA (150 μg/kg Monarsen) caused an increase in plasma AChE activity within 5 h.[29] This increase was effectively suppressed by a higher oral dose of 500 μg/kg of antisense and even more so by its intravenous administration, reflecting a dose-dependent prevention of peripheral AChE-R synthesis.

Following a 7-day course of daily oral drug administration, AChE-R mRNA was induced in lumbar spinal cord neurons. This induction was again suppressed by the Monarsen antisense treatment. Because antisense agents are not considered to penetrate the blood–brain barrier,[47] this effect on spinal neurons' AChE-R likely acts through an indirect mechanism (e.g., penetration of proinflammatory cytokines through the blood–brain barrier).

Lumbar sections from monkeys treated daily with orally delivered Monarsen depicted higher fractions of both large and medium-sized IL-1β-positive cell bodies compared to cells from naïve animals, suggesting stress-induced inflammatory response. Smaller fractions of IL-1β-labeled cell bodies were found in sections from animals treated by intravenous 500 μg/kg/day Monarsen compared with 150 μg/kg/day oral treatment. In accordance with this, AChE-R mRNA and IL-1β levels in medium-sized neurons exhibited a direct linear correlation. Similarly, IL-6 expression was significantly suppressed in medium-sized and large spinal neurons following treatment with 500 μg/kg Monarsen via intravenous compared to oral administration, reflecting a dose-dependent global reduction in these neuronal proinflammatory cytokines by Monarsen treatment.

These findings indicate that spinal cord interneurons respond to stress by overproducing AChE-R, limiting excessive cholinergic input into motoneurons and thus maintaining their adequate function under stressful conditions. Nevertheless, as these spinal interneurons carry the α7 nicotinic AChR,[48] AChE-R accumulation would be proinflammatory because it likely impairs the inhibition over proinflammatory cytokines release from these neurons, similar to its function in macrophages.[45,49]

Together, these antisense studies suggest the existence of a multileveled chain of events: following the acute phase of ACh release under stress, comes AChE-R induction, facilitating an inflammatory response. Intriguingly, systemic inflammation (e.g., that induced by bacterial lipopolysaccharide [LPS]), can upregulate proinflammatory cytokines, such as IL-1β within the rodent brain without penetrating the blood–brain barrier.[50–54] In some cases, plasma IL induction preceded the central inflammation.[51] Systemic administration of LPS or IL-1β further stimulates vagal sensory pathways, regulating central nervous system neuronal function.[55] Compatible with this, peripheral inhibition of plasma AChE activity by Monarsen treatment can reduce peripheral IL release via increased peripheral ACh levels, which in turn further reduces the cytokine effects on brain neurons. This pathway may explain the reduction of proinflammatory cytokines by Monarsen treatment without its blood–brain barrier penetration. Parallel findings in EN101-treated rats support this notion.[56]

Suppression of inflammation by antisense treatment carries additional advantage for treating autoimmune diseases, such as myasthenia gravis. In addition to alleviating the muscle fatigue by increasing available ACh for AChR activation, counteracting the autoimmune destruction of AChR must carry a benefit by suppressing the inflammatory load. This may indeed explain the improved survival and clinical status in the EAMG rats[27] (FIG. 1).

STRESS CONDITIONS IN MYASTHENIC PATIENTS AND POSSIBLE IMPLICATION FOR ANTISENSE THERAPEUTICS

The abundance of AChE-R in plasma of myasthenia patients is variable,[27] as it is in healthy volunteers.[20] This may possibly reflect fluctuations attributable to mild stressors, such as emotional upset, fever, and systemic illness.[57] Acute and rapid deterioration in myasthenic patients, into a condition termed *myasthenic crisis* may occur due to a severe infection, most often pneumonia,[58] and proceed within a matter of hours to respiratory failure and severe limb weakness.[23] In light of AChE-R induction following acute psychological as well as physical stress,[9] weakness fluctuations as well as the myasthenic crisis may, at least in part, be attributed to AChE-R induction. These conditions, which can be refractory to small molecule anti-AChE agents, may possibly be better controlled by the above-described antisense treatment.

Progressive myasthenic weakness often requires a second-line treatment with corticosteroids (prednisone), which reduce the immune attack on AChRs and induce a marked improvement in the majority of patients.[59] Steroid treatment for autoimmune disease is commonly initiated by high-dose pulse therapy. Nevertheless, initiation of high-dose treatment in myasthenic patients commonly induces a unique transient exacerbation of symptoms, with paradoxically increased weakness within the first week of treatment.[23] Patients who suffer from weakness of oropharyngeal or respiratory muscles may hence progress under this condition into a life-threatening airway or ventilatory crisis. Therefore, these patients are often hospitalized to start this treatment, with low and slowly incrementing doses. The glucocorticoid response element (GRE) in the ACHE promotor[60] may be causally involved in this acute reaction. Indeed, steroid treatment induces AChE gene expression and promotes alternative splicing, facilitating AChE-R accumulation in both neurons[14] and hematopoietic cells.[61] In this case as well, Monarsen treatment may retrieve the cholinergic balance that is impaired by such stress events or steroid treatment.

EFFICIENCY AND SAFETY OF MONARSEN TREATMENT

EN101 treatment was found to be similarly effective, at the same nanomolar dose, in mice,[16] rats,[27] and cynomolgus monkeys,[29] in both the intravenous and the oral administration routes.

The chemical protection employed at the 3′ region offers several advantages. First, 2′-o-methyl blockade does not form the chiral derivatives characteristic of phosphorothioate oligonucleotides.[62] Second, it improves intestinal permeability[63] while maintaining the oligonucleotide's capacity to recruit RNase H through its unprotected part and tightening the hybridization bonds through the 2′-o-methyl groups.[9] Third, breakdown of the protected 3′-end leaves behind a naked oligonucleotide, which is rapidly degraded. This avoids retention of shorter oligonucleotide chains that may participate in nonspecific hybridization events, with increasing incidence under gradual nucleolytic degradation.[63,64]

Potential toxicity of antisense treatment was tested on cynomolgus monkeys, during and following Monarsen applications. Body weight, food consumption, general locomotor behavior, electrocardiography and blood pressure, blood cell counts, prothrombin time, and standard blood chemistry were all normal, as were post mortem organ weights and hematoxylin–eosin staining of tissue sections from brain, heart, kidneys, liver, lungs, spinal cord, and stomach. No treatment-related toxicity or inflammatory effects not related to the subject of the study were detected.[29]

Antisense treatment against AChE-R expression is currently enrolled in a phase II clinical trial in myasthenic patients, provided at a daily single orally delivered dose, which effectively replaces multiple daily doses of the conventionally employed anti-AChE drug pyridostigmine (Argov *et al.*, in preparation). Patients appear to be free of the characteristic intestinal side effects caused by the nonselective suppression of AChE by the current small molecule drug. Monarsen thus offers potential advantages over conventional cholinesterase inhibitors with respect to dosing, specificity, side-effect profile, duration of efficacy, and treatment regimen.

OPEN QUESTIONS

The rapid, low dose, and long-lasting effects of orally delivered Monarsen raise several important questions. It is not yet clear how such low doses of partially protected oligonucleotide chains affect multiple tissues in the rapid time scale observed, or how the effect lasts as long as it does. Part of the explanation may involve the feedback loop associating AChE-R overproduction with inflammatory reactions, and *vice versa*; other origins of this surprising efficacy may be due to yet undisclosed aptamer reactions (e.g., with toll-like receptors, similar to CpG oligonucleotides[65]) or intracellular elicitation of a micro-RNA pathway.[66] Further studies would be required to elucidate the molecular cascade(s) through which EN101/Monarsen elicits its beneficial effects.

ACKNOWLEDGMENTS

This work was supported by the Eric Roland Center for Neurodegenerative Diseases at the Hebrew University of Jerusalem and by Ester Neurosciences, Ltd., Tel-Aviv, Israel (to H.S.).

REFERENCES

1. SILMAN, I. & J.L. SUSSMAN. 2005. Acetylcholinesterase: 'classical' and 'non-classical' functions and pharmacology. Curr. Opin. Pharmacol. **5:** 293–302.
2. IMPERATO, A., S. PUGLISI-ALLEGRA, P. CASOLINI & L. ANGELUCCI. 1991. Changes in brain dopamine and acetylcholine release during and following stress are independent of the pituitary-adrenocortical axis. Brain Res. **538:** 111–117.
3. ENNIS, M. & M.T. SHIPLEY. 1992. Tonic activation of locus coeruleus neurons by systemic or intracoerulear microinjection of an irreversible acetylcholinesterase inhibitor: increased discharge rate and induction of C-fos. Exp. Neurol. **118:** 164–177.
4. KAUFER, D., A. FRIEDMAN, S. SEIDMAN & H. SOREQ. 1998. Acute stress facilitates long-lasting changes in cholinergic gene expression. Nature **393:** 373–377.
5. LEV-LEHMAN, E., T. EVRON, R.S. BROIDE, *et al.* 2000. Synaptogenesis and myopathy under acetylcholinesterase overexpression. J. Mol. Neurosci. **14:** 93–105.
6. BAUSERO, P., M. SCHMITT, J.L. TOUSSAINT, *et al.*1993. Identification and analysis of the human choline acetyltransferase gene promoter. Neuroreport **4:** 287–290.
7. CERVINI, R., L. HOUHOU, P.F. PRADAT, *et al.* 1995. Specific vesicular acetylcholine transporter promoters lie within the first intron of the rat choline acetyltransferase gene. J. Biol. Chem. **270:** 24654–24657.
8. MESHORER, E., D. TOIBER, D. ZUREL, *et al.* 2004. Combinatorial complexity of 5′ alternative acetylcholinesterase transcripts and protein products. J. Biol. Chem. **279:** 29740–29751.
9. SOREQ, H. & S. SEIDMAN. 2001. Acetylcholinesterase—new roles for an old actor. Nat. Rev. Neurosci. **2:** 294–302.
10. LI, Y., S. CAMP, T.L. RACHINSKY, *et al.* 1991. Gene structure of mammalian acetylcholinesterase. Alternative exons dictate tissue-specific expression. J. Biol. Chem. **266:** 23083–23090.
11. MESHORER, E. & H. SOREQ. 2006. Virtues and woes of AChE alternative splicing in stress-related neuropathologies. Trends Neurosci. **29:** 216–224.
12. MESHORER, E., B. BRYK, D. TOIBER, *et al.* 2005. SC35 promotes sustainable stress-induced alternative splicing of neuronal acetylcholinesterase mRNA. Mol. Psychiatry. **10:** 985–997.
13. KAUFER, D. & H. SOREQ. 1999. Tracking cholinergic pathways from psychological and chemical stressors to variable neurodeterioration paradigms. Curr. Opin. Neurol. **12:** 739–743.
14. MESHORER, E., C. ERB, R. GAZIT, *et al.* 2002. Alternative splicing and neuritic mRNA translocation under long-term neuronal hypersensitivity. Science **295:** 508–512.
15. FARCHI, N., H. SOREQ & B. HOCHNER. 2003. Chronic acetylcholinesterase overexpression induces multilevelled aberrations in mouse neuromuscular physiology. J. Physiol. **546:** 165–173.
16. COHEN, O., C. ERB, D. GINZBERG, *et al.* 2002. Neuronal overexpression of "readthrough" acetylcholinesterase is associated with antisense-suppressible behavioral impairments. Mol. Psychiatry **7:** 874–885.
17. MCEWEN, B.S. 1999. Stress and hippocampal plasticity. Annu. Rev. Neurosci. **22:** 105–122.
18. SAPOLSKY, R.M., L.M. ROMERO & A.U. MUNCK. 2000. How do glucocorticoids influence stress responses? Integrating permissive, suppressive, stimulatory, and preparative actions. Endocr. Rev. **21:** 55–89.

19. MEZEY, G. & I. ROBBINS. 2001. Usefulness and validity of post-traumatic stress disorder as a psychiatric category. Br. Med. J. **323:** 561–563.
20. SKLAN, E.H., A. LOWENTHAL, M. KORNER, *et al.* 2004. Acetylcholinesterase/paraoxonase genotype and expression predict anxiety scores in health, risk factors, exercise training, and genetics study. Proc. Natl. Acad. Sci. USA **101:** 5512–5517.
21. GRIFMAN, M., N. GALYAM, S. SEIDMAN & H. SOREQ. 1998. Functional redundancy of acetylcholinesterase and neuroligin in mammalian neuritogenesis. Proc. Natl. Acad. Sci. USA **95:** 13935–13940.
22. DRACHMAN, D.B. 1994. Myasthenia gravis. N. Engl. J. Med. **330:** 1797–1810.
23. ROPPER, A.H. & R.H. BROWN. 2005. Adams and Victor's Principles of Neurology, 8th ed. McGraw-Hill. New York.
24. GRIFMAN, M. & H. SOREQ. 1997. Differentiation intensifies the susceptibility of pheochromocytoma cells to antisense oligodeoxynucleotide-dependent suppression of acetylcholinesterase activity. Antisense Nucleic Acid Drug Dev. **7:** 351–359.
25. GRISARU, D., E. LEV LEHMAN, M. SHAPIRA, *et al.*1999. Human osteogenesis involves differentiation-dependent increases in the morphogenically active 3′ alternative splicing variant of acetylcholinesterase. Mol. Cell. Biol. **19:** 788–795.
26. GALYAM, N., D. GRISARU, M. GRIFMAN, *et al.* 2001. Complex host cell responses to antisense suppression of ACHE gene expression. Antisense Nucleic Acid Drug Dev. **11:** 51–57.
27. BRENNER, T., Y. HAMRA-AMITAY, *et al.* 2003. The role of readthrough acetylcholinesterase in the pathophysiology of myasthenia gravis. Faseb J. **17:** 214–22.
28. DORI, A., J. COHEN, W.F. SILVERMAN, *et al.* 2005. Functional manipulations of acetylcholinesterase splice variants highlight alternative splicing contributions to murine neocortical development. Cereb. Cortex **15:** 419–430.
29. EVRON, T., L.B. MOYAL-SEGAL, N. LAMM, *et al.* 2005. RNA-targeted suppression of stress-induced allostasis in primate spinal cord neurons. Neurodegenerative Dis. **2:** 16–27.
30. BIRIKH, K.R., E.H. SKLAN, S. SHOHAM & H. SOREQ. 2003. Interaction of "readthrough" acetylcholinesterase with RACK1 and PKCbeta II correlates with intensified fear-induced conflict behavior. Proc. Natl. Acad. Sci. USA **100:** 283–288.
31. NIJHOLT, I., N. FARCHI, M. KYE, *et al.* 2004. Stress-induced alternative splicing of acetylcholinesterase results in enhanced fear memory and long-term potentiation. Mol. Psychiatry **9:** 174–183.
32. GRISARU, D., V. DEUTSCH, M. SHAPIRA, *et al.* 2001. ARP, a peptide derived from the stress-associated acetylcholinesterase variant, has hematopoietic growth promoting activities. Mol. Med. **7:** 93–105.
33. PERRY, C., E.H. SKLAN & H. SOREQ. 2004. CREB regulates AChE-R-induced proliferation of human glioblastoma cells. Neoplasia **6:** 279–286.
34. SOREQ, H., R. BEN AZIZ, C.A. PRODY, *et al.*1990. Molecular cloning and construction of the coding region for human acetylcholinesterase reveals a G + C-rich attenuating structure. Proc. Natl. Acad. Sci. USA **87:** 9688–9692.
35. CHAN, R.Y., F.A. ADATIA, A.M. KRUPA & B.J. JASMIN. 1998. Increased expression of acetylcholinesterase T and R transcripts during hematopoietic differentiation is accompanied by parallel elevations in the levels of their respective molecular forms. J. Biol. Chem. **273:** 9727–9733.

36. SCHWARZ, M., Y. LOEWENSTEIN LICHTENSTEIN, D. GLICK, *et al.* 1995. Successive organophosphate inhibition and oxime reactivation reveals distinct responses of recombinant human cholinesterase variants. Brain Res. Mol. Brain Res. **31:** 101–110.
37. RAY, D.E. & P.G. RICHARDS. 2001. The potential for toxic effects of chronic, low-dose exposure to organophosphates. Toxicol. Lett. **120:** 343–351.
38. HALEY, R.W., S. BILLECKE & B.N. LA DU. 1999. Association of low PON1 type Q (type A) arylesterase activity with neurologic symptom complexes in Gulf War veterans. Toxicol. Appl. Pharmacol. **157:** 227–233.
39. VINCENT, A. 1999. Immunology of the neuromuscular junction and presynaptic nerve terminal. Curr. Opin. Neurol. **12:** 545–551.
40. MONIA, B.P. 1997. First- and second-generation antisense oligonucleotide inhibitors targeted against human c-raf kinase. Ciba Found Symp. **209:** 107–119, discussion 119–23.
41. ENGEL, A.G., E.H. LAMBERT & T. SANTA. 1973. Study of long-term anticholinesterase therapy. Effects on neuromuscular transmission and on motor end-plate fine structure. Neurology **23:** 1273–1281.
42. HUDSON, C.S., J.E. RASH, T.N. TIEDT & E.X. ALBUQUERQUE. 1978. Neostigmine-induced alterations at the mammalian neuromuscular junction II. Ultrastructure. J. Pharmacol. Exp. Ther. **205:** 340–356.
43. CZURA, C.J. & K.J. TRACEY. 2005. Autonomic neural regulation of immunity. J. Intern. Med. **257:** 156–166.
44. TRACEY, K.J. 2002. The inflammatory reflex. Nature **420:** 853–859.
45. WANG, H., M. YU, M. OCHANI, *et al.* 2003. Nicotinic acetylcholine receptor alpha7 subunit is an essential regulator of inflammation. Nature **421:** 384–388.
46. LI, Y., L. LIU, J. KANG, *et al.* 2000. Neuronal-glial interactions mediated by interleukin-1 enhance neuronal acetylcholinesterase activity and mRNA expression. J. Neurosci. **20:** 149–155.
47. TAVITIAN, B., S. TERRAZZINO, B. KUHNAST, *et al.* 1998. In vivo imaging of oligonucleotides with positron emission tomography. Nat. Med. **4:** 467–471.
48. HELLSTROM-LINDAHL, E., O. GORBOUNOVA, A. SEIGER, *et al.*1998. Regional distribution of nicotinic receptors during prenatal development of human brain and spinal cord. Brain Res. Dev. Brain Res. **108:** 147–160.
49. SHYTLE, R.D., T. MORI, K. TOWNSEND, *et al.* 2004. Cholinergic modulation of microglial activation by alpha 7 nicotinic receptors. J. Neurochem. **89:** 337–343.
50. BAN, E., F. HAOUR & R. LENSTRA. 1992. Brain interleukin 1 gene expression induced by peripheral lipopolysaccharide administration. Cytokine **4:** 48–54.
51. QUAN, N., S.K. SUNDAR & J.M. WEISS. 1994. Induction of interleukin-1 in various brain regions after peripheral and central injections of lipopolysaccharide. J. Neuroimmunol. **49:** 125–134.
52. PITOSSI, F., A. DEL REY, A. KABIERSCH & H. BESEDOVSKY. 1997. Induction of cytokine transcripts in the central nervous system and pituitary following peripheral administration of endotoxin to mice. J. Neurosci. Res. **48:** 287–298.
53. NGUYEN, K.T., T. DEAK, S.M. OWENS, *et al.* 1998. Exposure to acute stress induces brain interleukin-1beta protein in the rat. J. Neurosci. **18:** 2239–2246.
54. TURRIN, N.P., D. GAYLE, S.E. ILYIN, *et al.* 2001. Pro-inflammatory and anti-inflammatory cytokine mRNA induction in the periphery and brain following intraperitoneal administration of bacterial lipopolysaccharide. Brain Res. Bull. **54:** 443–453.

55. Ek, M., M. Kurosawa, T. Lundeberg & A. Ericsson. 1998. Activation of vagal afferents after intravenous injection of interleukin-1beta: role of endogenous prostaglandins. J. Neurosci. **18:** 9471–9479.
56. Pollak, Y., A. Gilboa, O. Ben-Menachem, *et al.* 2005. Acetylcholinesterase inhibitors reduce brain and blood interleukin-1beta production. Ann. Neurol. **57:** 741–745.
57. Bradley, W.G., R.B. Daroff, G.M. Fenichel & J. Jankovic. 2004. Neurology in Clinical Practice. Elsevier. Philadelphia.
58. Thomas, C.E., S.A. Mayer, Y. Gungor, *et al.* 1997. Myasthenic crisis: clinical features, mortality, complications, and risk factors for prolonged intubation. Neurology **48:** 1253–1260.
59. Pascuzzi, R.M., H.B. Coslett & T.R. Johns. 1984. Long-term corticosteroid treatment of myasthenia gravis: report of 116 patients. Ann. Neurol. **15:** 291–298.
60. Shapira, M., I. Tur-Kaspa, L. Bosgraaf, *et al.* 2000. A transcription-activating polymorphism in the ACHE promoter associated with acute sensitivity to anti-acetylcholinesterases. Hum. Mol. Genet. **9:** 1273–1281.
61. Grisaru, D., M. Pick, C. Perry, *et al.* 2006. Hydrolytic and nonenzymatic functions of acetylcholinesterase comodulate hemopoietic stress responses. J. Immunol. **176:** 27–35.
62. Kanaori, K., S. Sakamoto, H. Yoshida, *et al.* 2004. Effect of phosphorothioate chirality on i-motif structure and stability. Biochemistry **43:** 5672–5679.
63. Geary, R.S., T.A. Watanabe, L. Truong, *et al.* 2001. Pharmacokinetic properties of 2′-O-(2-methoxyethyl)-modified oligonucleotide analogs in rats. J Pharmacol. Exp. Ther. **296:** 890–897.
64. Opalinska, J.B. & A.M. Gewirtz. 2002. Nucleic-acid therapeutics: basic principles and recent applications. Nat. Rev. Drug Discov. **1:** 503–514.
65. Krieg, A.M. 2004. Antitumor applications of stimulating toll-like receptor 9 with CpG oligodeoxynucleotides. Curr. Oncol. Rep. **6:** 88–95.
66. Guimaraes-Sternberg, C., A. Meerson, I. Shaked & H. Soreq. 2006. MicroRNA modulation of megakaryoblast fate involves cholinergic signaling. Leuk. Res. **30:** 583–595.

Characterization of Antisense Oligonucleotides Comprising 2′-Deoxy-2′-Fluoro-β-D-Arabinonucleic Acid (FANA)

Specificity, Potency, and Duration of Activity

NICOLAY FERRARI,[a] DENIS BERGERON,[a] ANNA-LISA TEDESCHI,[a] MARIA M. MANGOS,[a] LUC PAQUET,[a] PAOLO M. RENZI[a,b] AND MASAD J. DAMHA[c]

[a]*Topigen Pharmaceuticals Inc., Montreal, Quebec, Canada H1W 4A4*

[b]*CHUM Research Center, Notre-Dame Hospital, Montreal, Quebec, Canada H2L 4M1*

[c]*Department of Chemistry, Otto Maass Chemistry Building, McGill University, Montreal, Quebec, Canada H3A 2K6*

ABSTRACT: **Antisense oligonucleotides (AON) are being developed for a wide array of therapeutic applications. Significant improvements in their serum stability, target affinity, and safety profile have been achieved with the development of chemically modified oligonucleotides. Here, we compared 2′-deoxy-2′-fluoro-β-D-arabinonucleic acid (FANA)-containing AONs with phosphorothioate oligodeoxynucleotides (PS-DNA), 2′-O-methyl-RNA/DNA chimeras and short interfering RNAs (siRNA) with respect to their target knockdown efficacy, duration of action and resistance to nuclease degradation. Results show that two different configurations of FANA/DNA chimeras (altimers and gapmers) were found to have potent antisense activity. Specific target inhibition was observed with both FANA configurations with an estimated EC_{50} value comparable to that of an siRNA but 20-to 100-fold lower than the other commonly used AONs. Moreover, the FANA/DNA chimeras showed increased serum stability that was correlated with sustained antisense activity for up to 4 days. Taken together, these results indicate that chimeric FANA/DNA**

Address for correspondence: Dr. M.J. Damha, Department of Chemistry, Otto Maass Chemistry Building, McGill University, 801 Sherbrooke Street West, Montreal, Quebec, Canada H3A 2K6; Voice: 514-398-7552; fax: 514-398-3797.
e-mail: masad.damha@mcgill.ca

Ann. N.Y. Acad. Sci. 1082: 91–102 (2006).
doi: 10.1196/annals.1348.032

AONs are promising new tools for therapeutic gene silencing when increased potency and duration of action are required.

Keywords: antisense oligonucleotide; FANA; siRNA; luciferase

INTRODUCTION

Phosphorothioate antisense oligonucleotides (PS-DNA)[1] represent a powerful tool for specific gene silencing and many are being pursued as potential means to inhibit expression of clinically relevant genes in a wide array of therapeutic applications. Unfortunately, PS-DNA possess a relatively low binding affinity for target RNA that impacts on their potency in antisense drug development.

In recent years, the avenue of novel nucleotide modifications has significantly accelerated current progress aimed at improving the properties of AONs.[2] Chemical modifications including 2′-O-methyl (2′-OMe)-RNA/DNA[3] and 2′-O-methoxyethyl RNA/DNA gapmers,[4] locked nucleic acids (LNA),[5] and mixed phosphodiester (PO) and PS AONs,[6] to mention a few, have all shown increased stability of the AON/target RNA duplex and increased resistance to nuclease degradation.

A promising new chemical modification is the 2′-deoxy-2′-fluoro-β-D-arabinonucleic acid analog (FANA).[7–11] FANA represents the first example of a fully 2′-modified nucleic acid that has both high-affinity RNA binding and retains RNase H-compatible properties,[8] suggesting that FANA-modified AONs may prove highly useful for gene targeting applications. Indeed, PS-FANA/DNA gapmers were found to have exceptionally potent intracellular antisense activity compared to PS-DNA AONs.[12] In addition, this activity was shown to be sequence-specific and mediated by intracellular RNase H. Importantly, unlike 2′-O-methylribose chimeric compounds, the potencies of the PS-FANA/DNA gapmers (IC_{50} <10 nM) were not significantly limited by the length of the DNA core.[12]

Recently, RNA interference (RNAi) has emerged as a highly efficient gene silencing mechanism and short interfering RNAs (siRNA) have been found to be very useful molecules for gene knockdown.[13–17] Although the AON and siRNA technologies share many practical problems such as mRNA site selection and delivery, siRNA have gained popularity, mainly because of their sub-nanomolar efficacy range and duration of activity.[13] However, a growing body of evidence suggests that nonspecific (off-target) effects associated with siRNA and the need for chemical modifications for *in vivo* applications constitute hurdles for their use as therapeutic drugs.[14]

In the present study, we compared two configurations of PS-FANA/DNA chimeras to PS-DNA, ribose-modified AONs and siRNA with respect to their potency, duration of activity, specificity, and stability. This comparative study revealed that PS-FANA/DNA "gapmers" (i.e. PS-FANA-"gapmer") had potency and duration of action comparable to siRNAs.

MATERIALS AND METHODS

Oligonucleotide Synthesis

Fluoroarabinonucleoside monomers and FANA/DNA mixmers were synthesized as previously described.[8] Unmodified phosphorothioate (PS), 2′-OMe-RNA/DNA and PO/PS-RNA/DNA chimeras were obtained commercially from the University of Calgary DNA Synthesis Laboratory (Calgary, AB). All AONs were purified by anion-exchange high performance liquid chromatography (HPLC) followed by desalting (SepPak cartridges, Waters Limited, Mississauga, ON). Five siRNAs (Qiagen, Mississauga, ON) targeting different regions of the luciferase gene were screened and the most potent one (siRNA TOP5013) was used in the study. Refer to TABLE 1 for a list of all AONs and siRNA used in this study.

Cell Culture and Transfection

The HeLa X1/5 cell line was maintained as previously described.[12] For transfection, 1.0×10^5 cells/well in 24-well plates were exposed to 60 nM of AON or siRNA using Lipofectamine 2000 (Invitrogen, Burlington, ON) in a 1:2 ratio according to the manufacturer's recommendations. Dose-response studies were performed using 60 nM of AON or siRNA whereby the effective AON or siRNA was serially diluted with an unrelated AON or siRNA, respectively, reducing the concentration of active oligonucleotide while keeping the final concentration of AON or siRNA constant. Cell metabolic activity, as an indicator of cellular toxicity resulting from AON or siRNA transfection, was assessed using the Alamar Blue™ fluorimetric assay (Medicorp, Montreal, QC).

Luciferase Activity Assay

Luciferase activity assays were performed using the luciferase assay system (BD Bioscience, Mississauga, ON) according to the manufacturer's protocol. Luminescence was measured using a microplate luminometer (Luminoskan Ascent, Thermo LabSystem) immediately following addition of the luciferin substrate solution.

Determination of Luciferase Protein Levels

Aliquots of cell lysates were centrifuged and protein content was measured.

TABLE 1.

Name	Length (mer)	Sequence 5′ - 3′	Chemistry
TOP4005	18	PS - ata tcc ttg tcg tat ccc	PS-DNA
Mixmers			
TOP4002	18	PS - ATA TCC ttg tcg TAT CCC	PS-FANA-Gapmer
TOP4018	18	PS - *AUAU* cct tgt cgt a *UCCC*	PS-2′Ome RNA-DNA-10 gapmer
TOP4021	18	PS - ATA tcct TGT cgta TCCC	PS-FANA-Altimer
TOP4022	20	dN-5′p5′-AUAU-$c_sc_st_st_sg_st_sc_sg_st_sa_s$-UCCC-3′p3′ dN	PO/PS-RNA-DNA-10 gapmer
Controls			
TOP4002Mi	18	PS - ATA CCC-ttt tct-TAC CCC	PS-FANA-Gapmer
TOP4005Mi	18	PS - ata ccc ttt tct tac ccc	PS-DNA
TOP4018Mi	20	PS - *ATAC* cct ttt ctt a *CCCC*	PS-2′Ome RNA-DNA-10 gapmer
TOP4021Mi	18	PS - ATA -ccct -TTT -ctta -CCCC	PS-FANA-Altimer
TOP4022Mi	20	dN-5′p5′-ATAC-$c_sc_st_st_st_sc_st_st_sa_s$-CCCC-3′p3′ dN	PO/PS-RNA-DNA-10 gapmer
siRNA			
TOP5013	21	GCUUGAAGUCUUUAAUUAAtt ggCGAACUUCAGAAAUUAAUU	siRNA
TOP5014	21	GCAAUAUUUAAAUCGUAAAtt ggCGUUAUAAAUUUAGCAUUU	Control siRNA
TOP5026	21	AGGGAUACGACAAGGAUAUtt ttUCCCUCUGCUGUUCCUAUA	siRNA

Lowercase letters = DNA; bold uppercase letters = FANA; uppercase letters = RNA; upper case italics letters = 2′-OMe-RNA; dB = abasic deoxynucleoside; underlined letters = nucleotide mismatch; PS = phosphorothioate; PO/PS = mix phosphodiester and phosphorothioate (subscript s); 3′p3′ = 3′-3′ inverted PO linkage connection.

Equivalent amounts of total protein (25 μg) were resolved by SDS-PAGE, and then transferred to PVDF membrane. Luciferase protein levels were determined by immunoblot analysis using a horseradish peroxidase-conjugated IgG goat polyclonal anti-firefly-luciferase antibody (Cederlane, Hornby ON). Immunoblots were visualized using the BM Chemiluminescence Blotting Substrate (Roche, Laval, QC).

Luciferase mRNA Quantification

For luciferase mRNA quantification, cells were lysed in cell lysis buffer (Quantigene, Mississauga, ON) and luciferase mRNA levels determined using a *P. pyralis* luciferase probe set and the Quantigene Kit (Panomics, Fremont, CA) according to an optimized protocol. Luciferase mRNA levels were normalized to mRNA levels from the housekeeping gene β2m (Panomics).

Stability of Oligonucleotides in Human Serum

AONs (20 μg in 1 mL) were diluted in human serum and incubated at 37°C. Aliquots of 0.1 A_{260} units of AON (corresponding to 4–5 μg of each test substrate) were collected after 6, 24, 48, 72, and 96 h and frozen until analyzed. Samples were analyzed by anion-exchange HPLC on an Agilent 1100 system with multiwavelength detection using a linear gradient of 0 – 80% $LiClO_4$ (1 M) in water over 80 min to resolve the products. The HPLC column used was a Waters Protein-Pak DEAE-5PW (Waters Limited, Mississauga, ON). Under these conditions, all protein material could conveniently be resolved from intact oligonucleotide and product peaks, which eluted toward the end of the run. Degradation data from the acquired chromatograms were processed using ChemStation software as supplied by the manufacturer.

RESULTS

Comparison of Activity Between PS-FANA/DNA Chimeras, 2′-OMe-RNA/DNA Gapmers, and siRNA

The objective of this study was to compare two FANA designs (PS-FANA-Gapmer, TOP4002; PS-FANA-Altimer, TOP4021) to PS-DNA (TOP4005) and two chimeric 2′OMe-RNA "gapmers" (TOP4018 and 4022) with respect to their capacities to knock down the expression of luciferase mRNA (HeLa X1/5 cells) using our established protocols (TABLE 1).[12] It should be noted that although all gapmers were of identical sequence, both of the 2′-OMe-RNA gapmers evaluated contained 10 central DNA units, whereas the PS-FANA

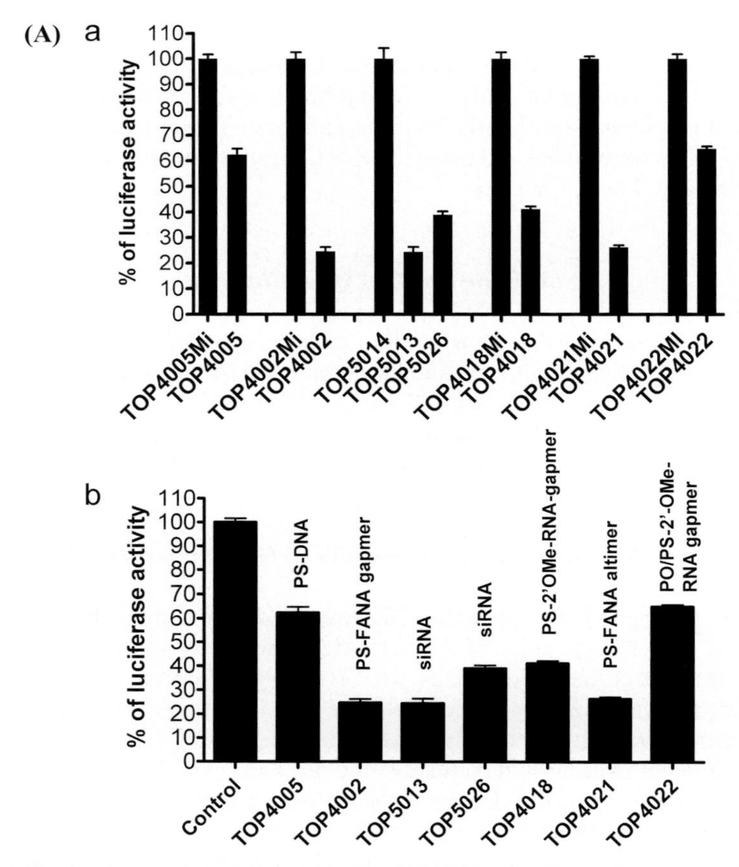

FIGURE 1. Potency of AON and siRNA at inhibiting luciferase in HeLa X1/5 cells. (**A**) Cells were transfected with 60 nM ONs and luciferase activity levels were measured 24 h post-transfection. (**B**) Dose responses were obtained for each AON and siRNA by transfecting cells with different amounts of active AON or siRNA for 24 h. Luciferase activity was measured and values normalized to the metabolic activity and compared to mismatch control AON or siRNA set at 100%. Data represent mean normalized luciferase activity ± SEM. Estimated EC_{50} values are indicated. (**C**) Luciferase protein levels were determined by western blot 24 h post-transfection with 60 nM of active AON or siRNA. Control cells remained untransfected or transfected with mismatch AON or siRNA.

gapmer contained only 6 such units. This was necessary in order to provide a more adequate and rational evaluation of these chemistries, since a previous study showed that these AONs gave the most optimal rate of RNase H-mediated mRNA cleavage ensuing intracellular activity (e.g., PS-2′-OMe-RNA gapmers with 6 DNA units showed little activity under the assay conditions).[12] Each antisense AON was designed to target the same sequence of the luciferase mRNA. In order to compare silencing efficiencies of AONs and siRNAs

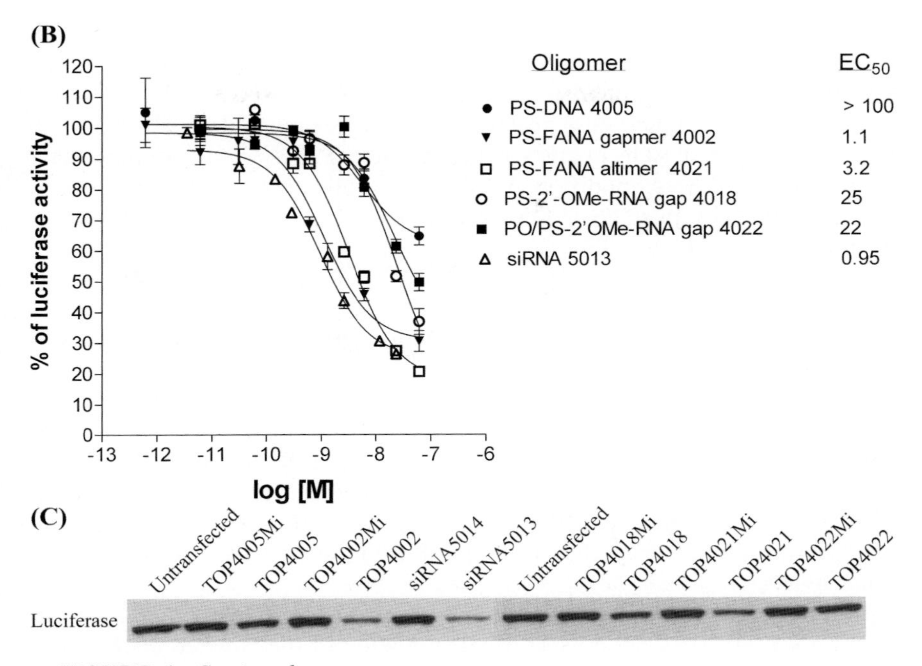

FIGURE 1. *Continued.*

directed against the same mRNA target, we also analyzed two siRNA duplexes. The first was isosequential to the AONs (TABLE 1; TOP5026), whereas the other (Qiagen) targeted a different mRNA region altogether.

Dose-response curves were generated for the PS-FANA gapmer (TOP4002), altimer (TOP4021), PS-DNA (TOP4005), PS-2′-OMe-RNA/DNA-gapmer, (TOP4018), and a PO/PS-RNA/DNA gapmer capped at its 5′- and 3′-terminus with inverted abasic nucleotide residues (GeneBloc TOP4022), and for siRNA TOP5013 (FIG. 1 A and B). Treatment with TOP4002 and TOP4021 resulted in a reduction of luciferase activity in a concentration-dependent manner with an estimated EC_{50} of 1.1 and 3.2 nM, respectively. The potency observed was comparable to that of siRNA TOP5013, which had an estimated EC_{50} of 0.95 nM in this system. The isosequential siRNA (TOP5026) showed potent inhibition, but its activity was less compared to the PS-FANA gapmer AON (FIG. 1 A). In addition, the data revealed an 8- to 20-fold increase in potency for both FANA gapmer (TOP4002) and altimer (TOP4021) as compared to the corresponding PS-2′-OMe-RNA/DNA and PO/PS-2′-OMe-RNA/DNA gapmer oligonucleotides (TOP4018 and TOP4022); they were also found to be approximatively 100-fold more potent than the all PS-DNA of identical sequence (TOP4005).

In order to determine whether the luciferase activity levels measured correlated with luciferase protein levels, Western blot analysis was performed

24 h after transfection with 60 nM of AON or siRNA. As shown in FIGURE 1C, the data show a relationship between luciferase activity and intracellular luciferase protein levels. Indeed, cells transfected with AON or siRNA had reduced luciferase protein levels when compared to cells exposed to mismatch control AON or siRNA. Taken together, these results suggest that, in this system, FANA-modified AONs have similar potency to siRNA and increased potency over PS-2′-OMe-RNA/DNA, PO/PS-2′-OMe-RNA/DNA gapmers and PS-DNA AONs.

Persistence of Activity of FANA-Containing Antisense Oligonucleotides

Next we examined the duration of gene silencing by TOP4002 and TOP4021. Sustained inhibition (~40%) of luciferase activity was observed in HeLa X1/5 cells at 96 h following exposure to TOP4002 and TOP4021 (FIG. 2 A). Exposure to the siRNA (TOP5013) also resulted in persistent inhibition (~80%) of luciferase activity at the 96-h time point. On the other hand, luciferase activity was almost completely restored 48 h following transfection with the other AONs.

In order to demonstrate that the observed effects of AON treatment on luciferase activity were due to specific degradation of the targeted mRNA, luciferase mRNA quantification was performed. Cells were lysed at different time points and target mRNA was quantified using the Quantigene method. FIGURE 2B shows that luciferase mRNA levels in cells exposed to the different AONs follow the same trend as the luciferase activity. However, in contrast to luciferase activity where maximum inhibition was observed at the 24 h time point, peak mRNA inhibition was measured at the 8-h time point. Again, TOP4002 showed increased activity over the other AONs with 60% target mRNA inhibition (8 h) compared to ~45% for TOP4018 and ~35% for TOP4005, TOP4021 and TOP4022. Taken together, these results indicate that, under these conditions, both FANA designs (gapmer and altimer) have a superior duration of activity over the other AONs. Moreover, the decreased luciferase activity closely correlated with reductions in mRNA levels, owing to the specific degradation of the target mRNA following antisense treatment.

Serum Stability

We next determined whether the greater efficacy and duration of activity observed for TOP4002 arose from higher resistance to nuclease degradation. The stability of the different chemistries over time was evaluated in human serum and degradation assessed by anion-exchange HPLC (FIG. 3). The results indicate that TOP4002 displayed increased resistance when compared to PS-DNA (TOP4005). In this assay, more than 95% of AON remained intact

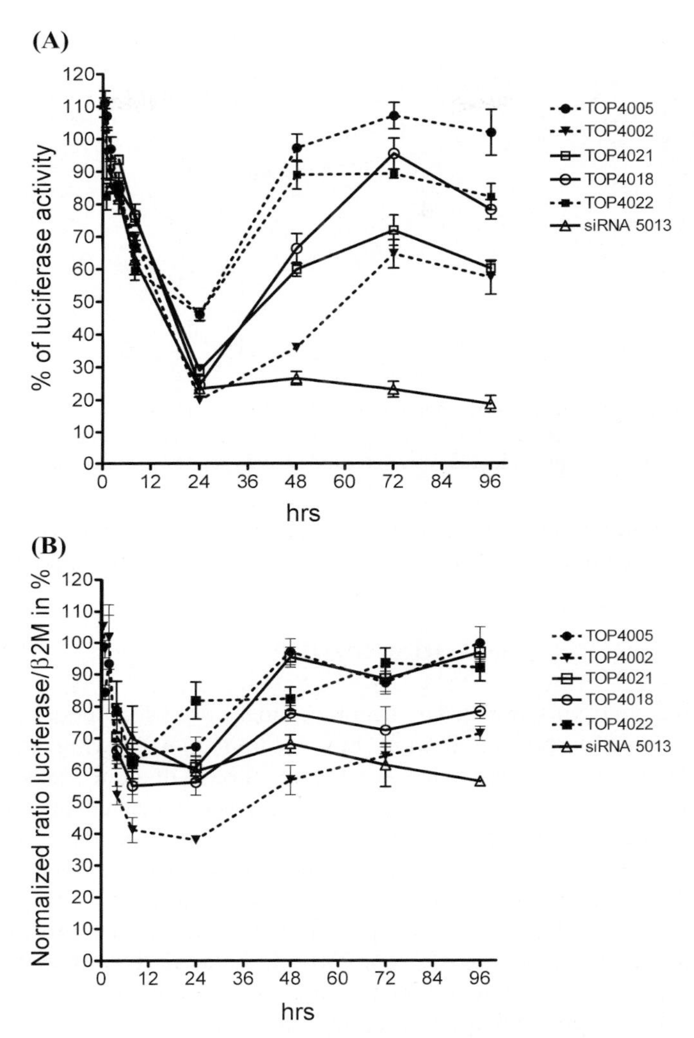

FIGURE 2. Time-course efficacy of different AONs and siRNAs targeting the luciferase mRNA in HeLa X1/5 cells. Cells were transfected with 60 nM of AON or siRNA. (**A**) Luciferase activity was measured 4, 8, 24, 48, 72, and 96 h post transfection. Data represent mean normalized luciferase activity ± SEM compared to mismatch control AON or siRNA set at 100%. (**B**) Luciferase mRNA levels were quantified at the same time points by Quantigene analysis (relative to β2m expression). Bars show mean luciferase/β2m ratios ± SEM.

after incubation with serum whereas only 45% of intact TOP4005 could be detected after 96 h, implying the PS-FANA gapmer species persist in cells and contribute in part to sustained activity on target mRNA suppression. Interestingly, TOP4018 showed similar nuclease resistance to TOP4002, despite its

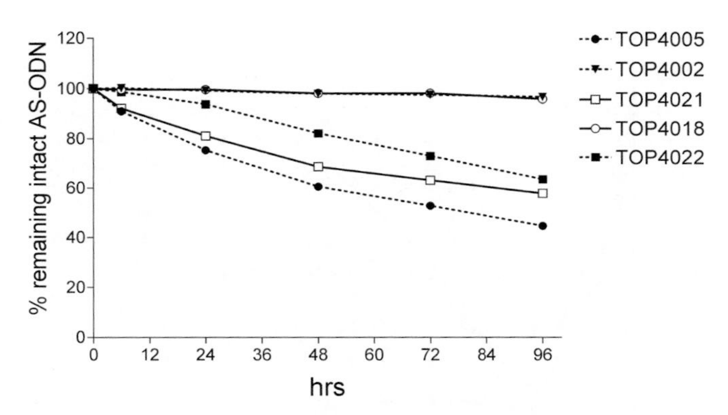

FIGURE 3. Stability of FANA-containing AON. AONs were incubated in human serum at 37°C and aliquots were taken at the time points indicated and analyzed by HPLC. The percentage of intact oligonucleotide was determined and graphed as a function of time.

generally poorer efficacy in luciferase suppression over 96 h (FIG. 2 B). On the other hand, both TOP4021 and TOP4022 displayed intermediate nuclease resistance.

DISCUSSION

Selective gene knockdown has become widely used for target validation and therapeutic purposes. We now have access to a growing arsenal of oligonucleotide-based approaches for the regulation of gene expression in mammalian cells.[15] The two most commonly exploited mechanisms by which short synthetic oligonucleotides can be used are RNase H-mediated degradation of the target mRNA and RISC-mediated mRNA cleavage (RNAi). Initial reports indicating siRNA to elicit potent target degradation of the complementary RNA sequence has fueled the appeal for antisense-based approaches.[16,17] The attraction for FANA and FANA/DNA chimeras derives from their nuclease resistance, and their ability to simultaneously increase the strength of AON:mRNA hybrids while triggering highly efficient RNase H-mediated degradation of target mRNA.[8,12]

The main objective of this study was to compare FANA-containing AONs to other chemically modified AONs and siRNAs. The potency of the antisense PS-FANA/DNA chimeras was at least 100-fold superior to that of a conventional PS-DNA antisense. The scrambled ONs of identical base content had no effect on luciferase expression, suggesting the observed inhibition to be sequence specific. Even more significantly, the low nM range of activity of the PS-FANA gapmer (EC_{50} ~1 nM) was very similar to the EC_{50} obtained for an optimized siRNA (Qiagen), and 25-fold lower than that of two 2′-OMe-RNA "gapmers" (FIG. 1). In addition to potency, the duration of action of the PS-FANA gapmer and altimer was markedly increased compared to PS-DNA and

PS-2′OMe-RNA gapmers, as monitored by luciferase activity levels (FIG. 2). Stronger inhibition of luciferase activity was observed for the PS-FANA-gapmer and siRNA ONs up to 4 days post-transfection. For the various chemistries tested, there was a general correlation between luciferase mRNA level and luciferase activity (FIG. 2B). Figure 2B shows that the luciferase mRNA levels were decreased by 50% 4 h after exposure with the antisense PS-FANA gapmer AON, and that strong inhibition continued up to 96 h likely due to the enhanced nuclease resistance of this AON. The correlation between mRNA levels and luciferase activity was less evident for the Qiagen siRNA (FIG. 2 B). In order to compare silencing efficiencies of AONs and siRNA directed against the same target site, we also analyzed an siRNA duplex that was isosequential to the AONs (FIG. 1A; TOP5026). This siRNA shows potent inhibition, but its activity was less compared to the PS-FANA gapmer AON (FIG. 1 A).

In summary, these experiments and others recently described[12,23] establish that PS-FANA gapmers and altimers serve as excellent models of antisense agents that have enhanced resistance to the action of degradative nucleases present in serum, bind to mRNA through duplex formation, elicit RNase H activity, and potently inhibit intracellular specific gene expression in a highly persistent manner. Accordingly, FANA-based AONs have potential utility as therapeutic agents and/or tools for the study and control of specific gene expression in cells and organisms.

ACKNOWLEDGMENTS

We acknowledge financial support by Topigen Pharmaceuticals and a CIHR grant to MJD. We thank Dr. Paul Wotton for encouragement and support.

REFERENCES

1. ECKSTEIN, F. 2000. Phosphorothioate oligodeoxynucleotides: what is their origin and what is unique about them? Antisense Nucleic Acid Drug Dev. **10:** 117–121.
2. KURRECK, J. 2003. Antisense technologies. Improvement through novel chemical modifications. Eur. J. Biochem. **270:** 1628–1644.
3. YOO, B.H. *et al.* 2004. 2′-O-methyl-modified phosphorothioate antisense oligonucleotides have reduced non-specific effects *in vitro*. Nucleic Acids Res. **32:** 2008–2016.
4. MCKAY, R.A. *et al.* 1999. Characterization of a potent and specific class of antisense oligonucleotide inhibitor of human protein kinase C-alpha expression. J. Biol. Chem. **274:** 1715–1722.
5. WAHLESTEDT, C. *et al.* 2000. Potent and nontoxic antisense oligonucleotides containing locked nucleic acids. Proc. Natl. Acad. Sci. USA **97:** 5633–5638.

6. STERNBERGER, M. *et al.* 2002. GeneBlocs are powerful tools to study and delineate signal transduction processes that regulate cell growth and transformation. Antisense Nucleic Acid Drug Dev. **12:** 131–143.
7. WILDS, C.J. & M.J. DAMHA. 1999. Duplex recognition by oligonucleotides containing 2′-deoxy-2′-fluoro-D-arabinose and 2′-deoxy-2′-fluoro-D-ribose. Intermolecular 2′-OH-phosphate contacts versus sugar puckering in the stabilization of triple-helical complexes. Bioconjug. Chem. **10:** 299–305.
8. WILDS, C.J. & M.J. DAMHA. 2000. 2′-Deoxy-2′-fluoro-beta-D-arabinonucleosides and oligonucleotides (2′F-ANA): synthesis and physicochemical studies. Nucleic Acid Res. **28** 3625–3635.
9. NORONHA, A.M. *et al.* 2000. Synthesis and biophysical properties of arabinonucleic acids (ANA): circular dichroic spectra, melting temperatures, and ribonuclease H susceptibility of ANA.RNA hybrid duplexes. Biochemistry **39:** 7050–7062.
10. TREMPE, J.F. *et al.* 2001. NMR solution structure of an oligonucleotide hairpin with a 2′F-ANA/RNA stem: implications for RNase H specificity toward DNA/RNA hybrid duplexes. J. Am. Chem. Soc. **123:** 4896–4903.
11. DAMHA, M.J. *et al.* 2001. Properties of arabinonucleic acids (ANA & 2′F-ANA): implications for the design of antisense therapeutics that invoke RNase H cleavage of RNA. Nucleosides Nucleotides Nucleic Acids **20:** 429–440.
12. LOK, C.C. *et al.* 2002. Potent gene-specific inhibitory properties of mixed-backbone antisense oligonucleotides comprised of 2′-deoxy-2′-fluoro-D-arabinose and 2′-deoxyribose nucleotides. Biochemistry **41:** 3457–3467.
13. ACHENBACH, T.V., B. BRUNNER & K. HEERMEIER. 2003. Oligonucleotide-based knockdown technologies: antisense versus RNA interference. Chembiochem **4:** 928–935.
14. SLEDZ, C.A. & B.R. WILLIAMS. 2004. RNA interference and double-stranded-RNA-activated pathways. Biochem. Soc. Trans. **32:** 952–956.
15. SCHERER, L.J. & J.J. ROSSI. 2003. Approaches for sequence-specific knockdown of mRNA. Nat. Biotechnol. **21:** 1457–1465.
16. ELBASHIR, S.M. *et al.* 2001. Duplexes of 21-nucleotide RNA s mediate RNA interference in cultured mammalian cells. Nature **411:** 494–498.
17. CAPLEN, N.J. *et al.* 2001. Specific inhibition of gene expression by small double-stranded RNA s in invertebrate and vertebrate systems. Proc. Natl. Acad. Sci. USA **98** 9742–9747.
18. GRUNWELLER, A. *et al.* 2003. Comparison of different antisense strategies in mammalian cells using locked nucleic acids, 2′-O-methyl RNA, phosphorothioates and small interfering RNA. Nucleic Acids Res. **31:** 3185–3193.
19. VICKERS, T.A. *et al.* 2003. Efficient reduction of target RNAs by small interfering RNA and RNase H-dependent antisense agents. J. Biol. Chem. **278:** 7108–7118.
20. BRAASCH, D.A. *et al.* 2002. Antisense inhibition of gene expression in cells by oligonucleotides incorporating locked nucleic acids: effect of mRNA target sequence and chimera design. Nucleic Acids Res. **30:** 5160–5167.
21. JEPSEN, J.S. & J. WENGEL. 2004. LNA-antisense rivals siRNA for gene silencing. Curr. Opin. Drug Discov. Dev. **7:** 188–194.
22. FILIPOWICZ, W. 2005. RNAi: the nuts and bolts of the RISC machine. Cell **122:** 17–20.
23. Kalota, A. *et al.* 2006. 2′-Deoxy-2′-fluoro-ß-D-arabinonucleic acid (2′F-ANA) modified oligonucleotides (ON) effect highly efficient, and persistent, gene silencing. Nucleic Acids Res. 34**:** 451–461.

Anti-HIV Activity of Steric Block Oligonucleotides

GABRIELA IVANOVA,[a] ANDREY A. ARZUMANOV,[a] JOHN J. TURNER,[a] SANDRINE REIGADAS,[b] JEAN-JACQUES TOULMÉ,[b] DOUGLAS E. BROWN,[c] ANDREW M.L. LEVER,[c] AND MICHAEL J. GAIT[a]

[a]*Medical Research Council, Laboratory of Molecular Biology, Cambridge CB2 2QH, UK*

[b]*Institut Européen de Chimie et Biologie, 33607 PESSAC, France*

[c]*Department of Medicine, University of Cambridge, Addenbrooke's Hospital, Cambridge CB2 2QQ, UK*

ABSTRACT: The unabated increase in spread of HIV infection worldwide has redoubled efforts to discover novel antiviral and virucidal agents that might be starting points for clinical development. Oligonucleotides and their analogs targeted to form complementary duplexes with highly conserved regions of the HIV RNA have shown significant antiviral activity, but to date clinical studies have been dominated by RNase H-inducing oligonucleotide analog phosphorothioates (GEM 91 and 92) that have specificity and efficacy limitations. However, they have proven the principle that oligonucleotides can be safe anti-HIV drugs. Newer oligonucleotide analogs are now available, which act as strong steric block agents of HIV RNA function. We describe our ongoing studies targeting the HIV-1 *trans*-activation responsive region (TAR) and the viral packaging signal (psi) with steric block oligonucleotides of varying chemistry and demonstrate their great potential for steric blocking of viral protein interactions *in vitro* and in cells and describe the first antiviral studies. Peptide nucleic acids (PNA) disulfide linked to cell-penetrating peptides (CPP) have been found to have particular promise for the lipid-free direct delivery into cultured cells and are excellent candidates for their development as antiviral and virucidal agents.

KEYWORDS: HIV; antiviral; virucide; oligonucleotide; steric block

In 2004, some 5 million new cases of HIV infection were reported worldwide (http://www.unaids.org/wad2004/EPIupdate2004_html_en/epi04_02_en.htm). In the absence of a widely applicable vaccine against HIV, the major chemical weapons available to combat HIV infection and AIDS are antiretroviral agents,

Address for correspondence: Michael J. Gait, Medical Research Council, Laboratory of Molecular Biology, Cambridge CB2 2QH, UK. Voice: +44-1223-248011; fax: +44-1223-402070.
e-mail: mgait@mrc-lmb.cam.ac.uk

Ann. N.Y. Acad. Sci. 1082: 103–115 (2006).
doi: 10.1196/annals.1348.033

which are generally used in combination (highly active anti-retroviral therapy [HAART]). Despite the development of many new antiretroviral agents,[1,2] there is still a pressing need to find more effective and better-tolerated anti-HIV agents, better drug combinations, and agents with virucidal activity. Topical virucides are in clinical trials but none have yet been clinically approved for the control of HIV spread.

Most anti-HIV drugs are small molecules targeted at the viral enzymes, such as reverse transcriptase (RT), protease, and integrase. However, many other molecular strategies are being developed aimed at interfering with virus replication.[3] While many of these newer strategies are aimed primarily at gene therapy,[4] it is unlikely that this will prove applicable to a large enough number of patients worldwide to meet the perceived needs. It is important therefore that intensive efforts continue to identify further HIV targets suitable for antiviral and virucidal drug development.

The genome of HIV is small: two identical RNA strands of approximately 9000 nucleotides are found in a virion that includes several viral proteins. Much of the RNA sequence is able to mutate rapidly, because the infidelity of the HIV RT during replication leads to point mutation and recombination, and thereby endows the virus with resistance to drug action. However, since molecular function is packed densely into the RNA sequence, a few functional regions have short sequences that are highly conserved between viral isolates and rarely or never mutated. These occur particularly often within the first 350 nucleotides known as the RNA leader sequence (FIG. 1). Synthetic oligonucleotides and their analogs that are exactly complementary to such conserved HIV RNA sequences, which are commonly essential to HIV replication, are obvious starting points for anti-HIV drug development. Unfortunately, the challenges of developing oligonucleotide-type reagents as anti-HIV therapeutics have taxed many scientific groups and companies.

A good historical example of the particular problems facing oligonucleotide drug development is that of the 26-mer oligodeoxynucleotide phosphorothioate (PS) GEM 91.[5,6] PS oligodeoxynucleotides, in which a nonbridging oxygen atom is replaced by sulfur (FIG. 2), show reasonable stability in serum and were found to be well tolerated in humans during phase I studies. GEM 91 showed good pharmacokinetics in monkeys.[7] Unfortunately, GEM 91 was withdrawn from phase II trials in 1997 due to insufficient clinical benefit and a reduction in platelet counts. The clinical failure of GEM 91 (and similar PS oligonucleotides in other companies) has a lot to do with the limitations imposed on the types of chemical modification that can be incorporated into the oligonucleotide backbone while maintaining the desired mode of action. It was believed that recognition of the duplex formed between the oligonucleotide and the RNA target by RNase H, and subsequent cleavage of the RNA strand, was essential to obtain sufficient biological efficacy. Such cleavage destroys the target RNA, thereby preventing gene expression or other RNA function. A PS backbone is one of the very few that permits such RNase H recognition and

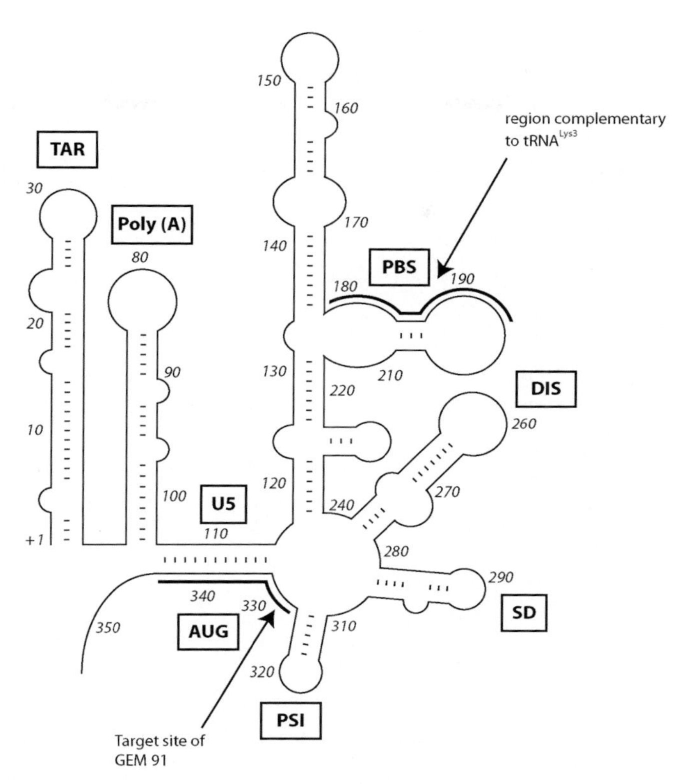

FIGURE 1. Secondary structure schematic model of the HIV-1 leader sequence showing the regions of functionality and the site of the clinically investigated PS oligonucleotide GEM 91.

was chemically straightforward to introduce. Surprisingly, much of the cellular anti-HIV activity of GEM 91 was later found to be due to nonsequence-specific effects, such as blocking of virus interaction with cells and inhibition of the RT enzyme.[8]

A second-generation version (GEM 92) was also developed, where the flanking sequences were substituted with 2′-*O*-methylribonucleotides (OMe) (FIG. 2) while maintaining a central core of PS deoxynucleotides. Introduction of the OMe units increases the binding to RNA, improves serum stability, and oral bioavailability, but the nonsequence-specific effects of all PS backbone are likely to remain. Unfortunately, GEM 92 development has not proceeded beyond phase I clinical trials. Another general problem that was identified with PS oligodeoxynucleotides is that certain sequences that include a Cp(s)G motif induce dramatic immune modulation as a result of the activation of certain toll-like receptors.[9] Although GEM 91 and 92 do not contain such a motif, some of the clinical responses of other antisense PS oligonucleotides are thought to be the result of immune stimulation.

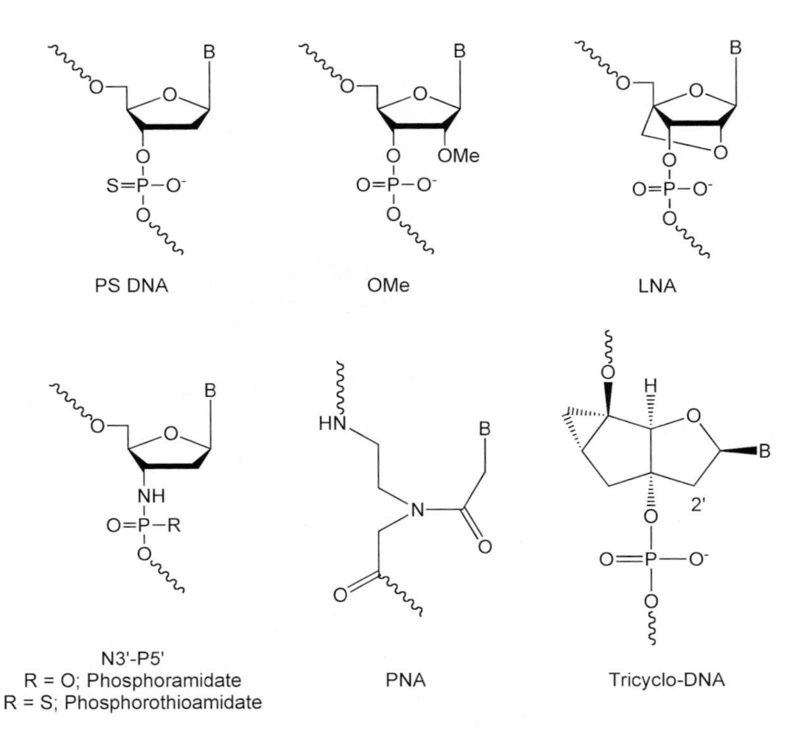

FIGURE 2. Structures of nucleotide analogs used in steric block and antisense oligonucleotide studies.

RNA interference has been much heralded as an anti-HIV strategy (recent reviews[10,11]). Much of the effort has concentrated on development of vectors to allow delivery of siRNA or shRNA into human progenitor cells for a gene therapy approach and also in finding susceptible sites within HIV RNA.[12,13] However, there are considerable problems with HIV escape from RNA interference[14] and it was shown recently that HIV-1 alters the structure of its RNA genome to evade RNAi effects.[15] It is likely that several siRNA or shRNAs will need to be engineered in tandem to reduce the chance of viral escape.

As a potential drug, siRNA needs to be heavily chemically modified in order to maintain sufficient stability in human circulation, since RNA is prone to very rapid degradation by nucleases. Recent results in blocking Hepatitis B Virus (HBV) replication in mice using heavily modified duplex siRNA formulated in a complex liposome coating appear promising, since the half-life in mouse circulation for the siRNA was 6.5 h,[16] suggesting that it might be possible to attempt to target HIV RNA *in vivo*. One advantage of siRNA is its extended time of intracellular activity. One obvious initial target would be the primer-binding site (PBS) where there are 18 residues of highly conserved sequence exactly complementary to the $tRNA^{Lys3}$ that acts as the primer for reverse transcription (FIG. 1). An unmodified siRNA duplex targeted to the PBS has already been shown to have considerable anti-HIV activity in cell culture.[12] So far no anti-HIV effects *in vivo* of modified siRNA have been reported.

STERIC BLOCK OLIGONUCLEOTIDES AND HIV RNA TARGETING: AN ALTERNATIVE CONCEPT

Both conventional antisense PS oligonucleotides and siRNA depend on the action of cellular enzymes (RNase H or the RISC complex, respectively) for activity. By contrast, it is possible to target RNA sequences that are essential for specific cellular function merely by forming a duplex between the RNA target and a synthetic oligonucleotide so that the function of the RNA (e.g., translation, splicing, polyadenylation, etc.) is sterically blocked.[17,18] In contrast to conventional antisense oligonucleotides, there is no requirement for RNase H recognition and thus steric block oligonucleotides can have any chemistry or added functionality that is compatible with good sequence-specific binding to the RNA target. Since binding to an incorrect site will be generally of no biological consequence, steric block oligonucleotides are much less likely to show off-target RNA inhibition effects. A disadvantage is that stoichiometric quantities, compared to the RNA target, need to be maintained intracellularly for activity. The ability to manipulate chemistry widely makes this class of oligonucleotides much more tractable to structure–activity experimentation, however.

Most of the important steric block oligonucleotides have targeted the HIV-1 leader RNA sequence, which has many regions essential to virus function, including the *trans*-activation responsive element (TAR), PBS, dimerization initiation site (DIS), packaging signal (psi), and translation initiator AUG, which, in one model, is thought to be complexed to a region of U5 close to the poly(A) stem-loop (FIG. 1). GEM 91 was targeted to the AUG region.[5] The viral leader sequence has been suggested to undergo various radical structural changes, for example to permit RNA dimerization,[19,20] but the first snapshots of the RNA structure in virions suggest that the structure is predominantly in a single conformation (FIG. 1).[21]

Steric blocking of HIV reverse transcription by oligodeoxyribonucleotides complementary to the region adjacent to the PBS (pre-PBS) or to the U5 region of the HIV RNA leader sequence was shown many years ago,[22] as well as for other retroviral reverse transcription.[23] Both oligonucleotides, when conjugated to poly-L-lysine, showed submicromolar inhibition of HIV infection in MOLT-4 cells in culture and blockage was shown to occur at the level of proviral DNA synthesis.[24]

Steric block 21-mer OMe/PS oligonucleotides targeted to the HIV PBS or splice acceptor site were shown to block the cytopathic activity of freshly infected MOLT-4 cells, but there was no activity in a chronically HIV-infected cell model.[25] The addition of PS linkages into a 2′-*O*-methyloligonucleotide backbone causes a substantial drop in binding strength to structured RNAs (for example, for 17-mers to TAR),[26] which may explain why HIV inhibitory activity has not been reported with PS/OMe oligonucleotides targeting the HIV leader, in which there is considerable secondary structure. OMe oligonucleotides without PS linkages have insufficient stability to serum nucleases, and therefore cannot be considered for therapeutic use. Nevertheless, 20-mer 2′-*O*-methyloligonucleotides were shown to act as competitive inhibitors of $tRNA^{Lys3}$ in binding to the PBS and to inhibit HIV infectivity in a HeLa P4 cell line expressing CD4 receptors and the *lacZ* gene under the control of HIV-1 LTR.[27]

The highly conserved apical region of the 59-residue TAR stem-loop is a particularly good site for targeting by steric block oligonucleotides (FIG. 3 A).[28] TAR is sited at the extreme 5′-end of the viral RNA and acts as a binding site for the HIV *trans*-activator protein Tat together with host cellular factors to trigger a large boost in transcriptional elongation by the transcription complex.[29,30] TAR also plays an important role in initiation of reverse transcription.[31] Tat *trans*-activation is essential for viral replication, but so far no small molecule inhibitors have proved to be clinical anti-HIV candidates.[32] Several years ago, three types of chemically modified 16-mer oligonucleotides (2′-*O*-methyl, N3′-P5′-phosphoramidate, and peptide nucleic acid [PNA]) (FIG. 2) targeted to TAR were shown to be efficient and sequence-specific inhibitors of HIV reverse transcription with IC_{50} in the nM range.[33] In contrast to PS oligonucleotides, they did not bind nonspecifically to HIV RT.

Our own studies have focused in particular on the inhibition of Tat-dependent *trans*-activation by targeting the apical HIV-1 TAR stem-loop (FIG. 3 A). We showed first that 12-mer oligonucleotides containing either all OMe nucleotides, a mixture of OMe and locked nucleic acid (LNA) units linkages (FIG. 2), or a PNA oligomer inhibited *in vitro* HIV-1 Tat binding and Tat-dependent *in vitro* transcription sequence specifically.[34,35] We then showed that a 16-mer OMe/LNA mixmer sequence specifically inhibited Tat-dependent *trans*-activation in a HeLa cell reporter system involving stably integrated luciferase plasmids with IC_{50} of 120 nM, when delivered by cationic lipids.[36,37] The LNA units were essential to the activity and we showed by use of a fluorescently labeled OMe/LNA oligonucleotide that this was entirely due to substantially improved cellular uptake and localization within the cell nucleus.[37] Probably the added hydrophobicity of the LNA units helps in packaging of the oligonucleotide by cationic lipids, since we showed that replacement of the LNA units by isomeric α-LNA units, which have completely different stereochemistry, did not reduce the activity, thus ruling out conformational effects.[37] Interestingly, a 12-mer OMe/LNA mixmer was the minimal length for cellular activity, whereas a 9-mer was active in blocking Tat-dependent *in vitro* transcription, pointing to the requirement for oligonucleotides to be maintained on the RNA target for a significant length of time to elicit steric block activity inside cells.[37]

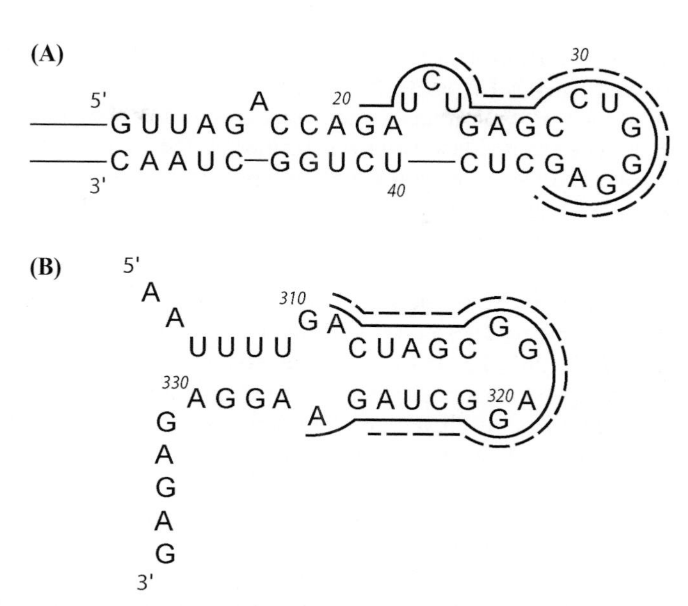

FIGURE 3. (**A**) Secondary structure of the apical part of the RNA stem-loop known as TAR, binding site of the HIV Tat protein, showing the region targeted by a 16-mer oligonucleotide (*solid line*) and 12-mer (*dashed line*) used in our studies. Numbering refers to the residues counted from the 5′-end of HIV-1 transcripts. (**B**) Secondary structure of the SL3 RNA stem-loop that is the binding site for HIV-1 Gag protein, showing the regions targeted by a 16-mer oligonucleotide (*solid line*) and a 14-mer (*dashed line*) used in our studies.

Very recently we have been exploring a second target for synthetic olignucleotide analogs on the HIV-1 RNA leader, the viral packaging signal psi. The 14-mer and 16-mer OMe and OMe/LNA mixmer oligonucleotides targeted to the well-conserved SL3 stem-loop (FIG. 3 B) are able to block the binding of HIV-1 Gag protein sequence specifically, thus demonstrating that SL3 is the major determinant of Gag binding (Brown, D.E., Joy, E., Greatorex, J., Gait, M.J. & Lever, A.M.L., manuscript submitted).

A key issue is to what extent the chemistry of the oligonucleotide backbone, once delivered into cells in culture by cationic lipids, can subsequently affect the ability of HIV-1 to replicate within those cells. Such activity could be obtained in this case either by blocking the ability of the HIV RNA to be reverse transcribed or to express full length RNA from an integrated provirus. We have shown recently that a OMe/LNA 16-mer targeted to the TAR loop (FIG. 3 A) and OMe/LNA 14-mer or16-mer oligonucleotides targeted to the SL3 loop (FIG. 3 B), when delivered by Lipofectamine2000, showed in each case sequence-specific inhibition of syncitia formation induced by HIV-1 infection in a HeLa T4 (P4) LTR-β-galactosidase (MAGI cell) model.[38] The 16-mer

OMe/LNA oligonucleotides targeted to either target inhibited sequence specifically the release of HIV-1 from Jurkat T cells.[39]

In a separate set of experiments, the levels of inhibition of expression of HIV-1 induced β-galactosidase were determined in the MAGI cell assay for 16-mers of several different oligonucleotide types. In preliminary studies, we have confirmed strong inhibitory activity of the 16-mer OMe/LNA as well as of a 16-mer N3′-P5′ phosphorothioamidate (FIG. 2) targeted to TAR to viral replication, while a N3′-P5′ phosphoramidate and a tricyclooligonucleotide had weak inhibitory activity when delivered into the cells by Lipofectamine2000 and subsequently infected with HIV-1 (Reigadas S., Ivanova, G., Arzumanov, A., Toulmé, J.-J. & Gait M.J., unpublished results).

Other research groups have targeted the TAR apical stem-loop with steric block oligonucleotides of different chemistry. Some inhibition of *trans*-activation in a transient cell reporter assay has been obtained with OMe oligonucleotides targeted to TAR that contain some methylphosphonate linkages.[40] More effective has been the use of PNA oligonucleotides originally aimed at blocking reverse transcription.[33,41] Inhibition of Tat-dependent *trans*-activation in cell culture with transiently transfected plasmids was achieved by electroporation of a 16-mer PNA into cells.[42] Interestingly, HIV replication in chronically infected H9 cells could also be achieved in the same way for the 16-mer but shorter 12-mer and 13-mer PNAs were much less effective.[43]

Another similar steric blocking approach suitable for use with apical loops of conserved regions of RNA structures, such as the TAR region of HIV-1, is the selection of oligonucleotide aptamers that form sequence-specific “kissing complexes” with the loops.[44,45] Such aptamers have remarkable sequence specificity, which is higher than that of antisense. Oligonucleotide aptamers that contain N3′-P5′-phosphoramidates[46] or DNA/LNA mixtures[47] inhibit Tat-dependent *in vitro* transcription, and cellular activity for such nuclease-stabilized aptamers is expected.

CELLULAR DELIVERY OF OLIGONUCLEOTIDES AND ANALOGS

Key to the possible use of steric block oligonucleotides as anti-HIV agents is to be able to enhance cellular uptake and delivery in the absence of transfection agents, such as cationic lipids. Although lipid formulation has recently been used successfully for anti-HBV modified siRNA delivery into mice,[16] it is not clear to what extent such formulation can be used for systemic use in humans to treat HIV infection. So far it has proved very difficult to obtain strong cellular delivery for oligonucleotide analogues into a wide range of cells in culture without the use of a transfection agent. Some improvements to lipid-free cell uptake have been found by reducing the overall charge of the oligonucleotide backbone, for example, by attaching cationic charges to the internucleotide phosphate linkages of α-oligonucleotides[48] or by use of hydrophobic thioester pro-oligonucleotide functions.[49] However, electrically neutral PNA and phosphorodiamidate morpholino (PMO) oligonucleotides are still not taken up significantly by cells, suggesting that lack of negative charge is not of itself sufficient to promote strong cellular delivery.

Some years ago it was found that cationic poly-L-lysine conjugates of oligonucleotides were delivered into HIV-infected cells.[24] More recently, a number of cell-penetrating peptides (CPP) have been identified and it was suggested that conjugation to oligonucleotides and their analogs might enhance cellular delivery (reviews[50–52]). Only one clear example of intracellular activity has been documented for CPP conjugates of a negatively charged steric block oligonucleotide in the absence of cationic lipids for the regulation of splicing.[53] By contrast, we found that several CPP conjugates of a fluorescein-labeled, 12-mer OMe/LNA oligonucleotide targeted to TAR were trapped in endosomal compartments of HeLa cells following free delivery and thus were unable to inhibit Tat-dependent *trans*-activation in a HeLa cell assay involving stably integrated luciferase plasmids.[54]

By contrast, we have found recently that certain CPPs (for example, Transportan or a cationic double CPP, R_6-Penetratin) disulphide-linked to a 16-mer PNA targeted to TAR were able to elicit significant inhibition of Tat-dependent *trans*-activation during lipid-free delivery for 24 h in our HeLa cell assay.[55] We found that fluorescently labeled CPP-PNA conjugates were trapped mostly in endosomal or membrane-bound compartments and their release (and subsequent *trans*-activation inhibition activity) could be enhanced by coadministration of the lysosomotropic agent chloroquine.

Meanwhile, another research group has shown that a disulphide-linked conjugate of 16-mer PNA with the CPP Transportan was able to elicit substantial anti-HIV activity in chronically infected CEM or Jurkat cells.[56] Further, this Transportan-PNA conjugate, in addition to its antiviral activity (IC_{50} 400 nM), was found to have very high activity as a virucidal agent (IC_{50} 66 nM) by pretreatment of HIV-1 pseudotype virions before cell infection.[57] At least part of this activity may be due to inhibition of reverse transcription within virions. Similar levels of virucidal activity were found for several other different CPP conjugates of the 16-mer PNA, but their antiviral activities were lower and varied significantly from 700–1100 nM IC_{50}.[58]

It is not clear as yet to what extent antiviral or virucidal activity can be maintained during extended virus challenge for cells treated with oligonucleotides targeting the TAR RNA region. There are also other regions of the HIV-1 RNA leader (U5, PBS, PSI, and AUG) that have excellent sequence conservation between virus strains and comparison of long-term virus inhibition for CPP-PNA or oligonucleotides of other chemistry targeted to each of these regions would be important in order to establish the optimal target site(s) for a steric block approach. Further, there is great scope to investigate other CPP conjugates of PNA that might have greater cell uptake or virion entry properties. Pharmacokinetics of nondisulfide-linked PNA peptides looks promising,[59] but in general very few *in vivo* studies of most newer oligonucleotide analogs are available as yet. In principle, it should also be possible to manipulate the pharmacology of steric block oligonucleotides by the attachment of other moieties that help retention in the human circulatory system (e.g., the recent example of cholesterol attachment to siRNA[60]). It should be noted that a steric block N3′-P5′-thiophosphoramidate oligonucleotide carrying a lipid tail is now entering clinical trials as an anticancer agent.[61]

Anti-HIV agents that have both antiviral and virucudal properties are badly needed. The promising recent cellular results with steric block oligonucleotides suggests it is now time for the approach to be taken more seriously and for some indepth virological and pharmacological studies to be carried out in order to determine the prospects for development of a steric block anti-HIV oligonucleotide as a clinical candidate.

ACKNOWLEDGMENTS

This work is funded in part by a grant from EC Framework 5 (contract QLK3-CT-2002-01989). Work on chloroquine enhancement of endosome release of PNA-peptides is carried out in collaboration with B. Lebleu and S. Abes (University of Montpellier).

REFERENCES

1. De Clercq, E. 2004. HIV-chemotherapy and -prophylaxis: new drugs, leads and approaches. Int. J. Biochem. Cell Biol. **36:** 1800–1822.
2. De Clercq, E. 2005. Emerging anti-HIV drugs. Expert Opin. Emerg. Drugs **10:** 241–274.
3. Nielsen, M.H., F.S. Pedersen & J. Kjems. 2005. Molecular strategies to inhibit HIV-1 replication. Retrovirology **2:** 10.
4. Strayer, D.S. *et al.* 2005. Current status of gene therapy strategies to treat HIV/AIDS. Mol. Ther. **11:** 823–842.
5. Lisziewicz, J. *et al.* 1994. Antisense oligodeoxynucleotide phosphorothioate complementary to Gag mRNA blocks replication of human immunodeficiency virus type 1 in human peripheral blood cells. Proc. Natl. Acad. Sci. USA **91:** 7942–7946.
6. Agrawal, S. & J.Y. Tang. 1992. GEM 91–An antisense oligonucleotide phosphorothioate as a therapeutic agent for AIDS. Antisense Res. Dev. **2:** 261–266.
7. Grindel, J.M. *et al.* 1998. Pharmacokinetics and metabolism of an oligodeoxynucleotide phosphorothioate (GEM 91) in cynomologous monkeys following intravenous infusion. Antisense Nucleic Acid Drug Dev. **8:** 43–52.

8. YAMAGUCHI, K. *et al.* 1997. The multiple inhibitory mechanisms of GEM 91, a gag antisense phosphorothioate oligonucleotide, for human immunodeficiency virus type 1. AIDS Res. Hum. Retroviruses **13:** 545–554.
9. AGRAWAL, S. & E.R. KANDIMELLA. 2004. Antisense and siRNA as agonists of toll-like receptors. Nat. Biotechnol, **22:** 1533–1537.
10. CULLEN, B.R. 2005. Does RNA interference have a future as a treatment for HIV-1 induced disease? AIDS Rev. **7:** 22–25.
11. TAKAKU, H. 2004. Gene silencing of HIV-1 by RNA interference. Antiviral Chem. Chemother. **15:** 57–65.
12. HAN, W. *et al.* 2004. Inhibition of human immunodeficiency virus type 1 replication by siRNA targeted to the highly conserved primer binding site. Virology **330:** 221–232.
13. CHANG, L.-J., X. LIU & J. HE. 2005. Lentiviral siRNAs targeting multiple highly conserved RNA sequences of human immunodeficiency virus type 1. Gene Ther. **12:** 1133–1144.
14. BODEN, D. *et al.* 2003. Human immunodeficiency virus type 1 escape from RNA interference. J. Virol. **77:** 11531–11535.
15. WESTERHOUT, E.M. *et al.* 2005. HIV-1 can escape from RNA interference by evolving an alternative structure in its RNA genome. Nucleic Acids Res. **33:** 796–804.
16. MORRISSEY, D.V. *et al.* 2005. Potent and persistent in vivo anti-HBV activity of chemically modified sirRNAs. Nat. Biotechnol. **23:** 1002–1007.
17. BAKER, B.F. *et al.* 1997. 2′-O-(2-methoxy)ethyl-modified anti-intercellular adhesion molecule 1 (ICAM-1) oligonucleotides selectively increase the ICAM-1 mRNA level and inhibit formation of the ICAM-1 translation initiation complex in human umbilical vein endothelial cells. J. Biol. Chem. **272:** 11994–12000.
18. KOLE, R., M. VACEK & T. WILLIAMS. 2004. Modification of alternative splicing by antisense therapeutics. Oligonucleotides **14:** 65–74.
19. HUTHOFF, H. & B. BERKHOUT. 2001. Two alternating structures of the HIV-1 leader RNA. RNA **7:** 143–157.
20. ABBINK, T.E.M. *et al.* 2005. The HIV-1 leader RNA conformational switch regulates RNA dimerization but does not regulate mRNA translation. Biochemistry **44:** 9058–9066.
21. PAILLART, J.C. *et al.* 2004. First snapshots of the HIV-1 RNA structure in infected cells and in virions. J. Biol. Chem. **279:** 48397–48403.
22. BORDIER, B. *et al.* 1992. In vitro effect of antisense oligonucleotides on human immunodeficiency virus type 1 reverse transcription. Nucleic Acids Res. **20:** 5999–6006.
23. BOIZIAU, C., N.T. THUONG & J.J. TOULMÉ. 1992. Mechanisms of the inhibition of reverse transcription by antisense oligonucleotides. Proc. Natl. Acad. Sci. USA **89:** 768–772.
24. BORDIER, B. *et al.* 1995. Sequence-specific inhibition of human immunodeficiency virus (HIV) reverse transcription by antisense oligonucleotides: comparative study in cell-free assays and in HIV-infected cells. Proc. Natl. Acad. Sci. USA **92:** 9383–9387.
25. SHIBAHARA, S. *et al.* 1989. Inhibition of human immunodeficiency virus (HIV-1) replication by synthetic oligo-RNA derivatives. Nucleic Acids Res. **17:** 239–252.
26. ECKER, D.J. *et al.* 1992. Pseudo-half knot formation with RNA. Science **257:** 958–961.

27. FREUND, F. *et al.* 2001. Inhibition of HIV-1 replication in vitro and in human infected cells by modified antisense oligonucleotides targeting the tRNA^{Lys3}/RNA initiation complex. Antisense Nucleic Acid Drug Dev. **11:** 301–315.
28. VICKERS, T.A. *et al.* 1991. Inhibition of HIV-LTR gene expression by oligonucleotides targeted to the TAR element. Nucleic Acids Res. **19:** 3359–3368.
29. KARN, J. 1999. Tackling Tat. J. Mol. Biol. **293:** 235–254.
30. RANA, T.M. & K.T. JEANG. 1999. Biochemical and functional interactions between HIV-1 Tat protein and TAR RNA. Arch. Biochem. Biophys. **365:** 175–185.
31. HARRICH, D. *et al.* 1997. Tat is required for efficient HIV-1 reverse transcription. EMBO J. **16:** 1224–1235.
32. KREBS, A. *et al.* 2003. Targeting the HIV trans-activation responsive region-approaches towards RNA-binding drugs. Chembiochem **4:** 972–978.
33. BOULMÉ, F. *et al.* 1998. Modified (PNA, 2′-O-methyl and phosphoramidate) anti-TAR antisense oligonucleotides as strong and specific inhibitors of in vitro HIV-1 reverse transcriptase. Nucleic Acids Res. **26:** 5492–5500.
34. ARZUMANOV, A. & M.J. GAIT. 1999. Inhibition of the HIV-1 tat protein-TAR RNA interaction by 2′-*O*-methyl oligoribonucleotides. *In* Collection Symposium Series, Vol. 2. A. Holy & M. Hocek, Eds.: 168–174 Academy of Sciences of the Czech Republic. Prague.
35. ARZUMANOV, A. *et al.* 2001. Oligonucleotide analogue interference with the HIV-1 Tat protein-TAR RNA interaction. Nucleosides Nucleotides Nucleic Acids **20:** 471–480.
36. ARZUMANOV, A. *et al.* 2001. Inhibition of HIV-1 Tat-dependent *trans*-activation by steric block chimeric 2′-O-methyl/LNA oligoribonucleotides. Biochemistry **40:** 14645–14654.
37. ARZUMANOV, A. *et al.* 2003. A structure-activity study of the inhibition of HIV-1 Tat-dependent *trans*-activation by mixmer 2′-O-methyl oligoribonucleotides containing locked nucleic acid (LNA), α-LNA or 2′-thio-LNA residues. Oligonucleotides **13:** 435–453.
38. BROWN, D. *et al.* 2005. Antiviral activity of steric-block oligonucleotides targeting the HIV-1 trans-activation response and packaging signal stem-loop RNAs. Nucleosides Nucleotides Nucleic Acids **24:** 393–396.
39. BROWN, D.E. *et al.* 2006. Inhibition of HIV-1 replication by oligonucleotide analogues directed to the packaging signal and trans-activation response region. Antivir. Chem. Chemother. **17:** 1–9.
40. HAMMA, T. *et al.* 2003. Inhibition of HIV Tat-TAR interactions by an antisense oligo-2′-O-methylribonucleoside methylphosphonate. Bioorg. Med. Chem. Lett. **13:** 1845–1848.
41. LEE, R. *et al.* 1998. Polyamide nucleic acid targeted to the primer binding site of the HIV-1 RNA genome blocks in vitro HIV-1 reverse transcription. Biochemistry **37:** 900–910.
42. MAYHOOD, T. *et al.* 2000. Inhibition of Tat-mediated transactivation of HIV-1 LTR transcription by polyamide nucleic acid targeted to the TAR hairpin element. Biochemistry **39:** 11532–11539.
43. KAUSHIK, M.J., A. BASU & P.K. PANDEY. 2002. Inhibition of HIV-1 replication by anti-transactivation responsive polyamide nucleotide analog. Antiviral Res. **56:** 13–27.
44. DUCONGÉ, F. & J.-J. TOULMÉ. 1999. In vitro selection identifies key determinants for loop-loop interactions: RNA apatamers selective for the TAR RNA element of HIV-1. RNA **5:** 1605–1624.

45. Boiziau, C. *et al.* 1999. DNA aptamers selected against the HIV-1 trans-activation-responsive RNA element form RNA-DNA kissing complexes. J. Biol. Chem. **274:** 12730–12737.
46. Darfeuille, F. *et al.* 2002. Loop-loop interaction of HIV-1 TAR RNA with N3′-5′ deoxyphosphoramidate aptamers inhibits *in vitro* Tat-mediated transcription. Proc. Natl. Acad. Sci. USA **99:** 9709–9714.
47. Darfeuille, F. *et al.* 2004. LNA/DNA chimeric oligomers mimic RNA aptamers targeted to the TAR RNA element of HIV-1. Nucleic Acids Res. **32:** 3101–3107.
48. Michel, T. *et al.* 2003. Cationic phosphoramidate a-oligonucleotides efficiently target single-stranded DNA and RNA and inhibit hepatitis C virus IRES-mediated translation. Nucleic Acids Res. **31:** 5282–5290.
49. Bologna, J.-C. *et al.* 2002. Uptake and quantification of intracellular concentration of lipophilic pro-oligonucleotides in HeLa cells. Antisense Nucleic Acid Drug Dev. **12:** 33–41.
50. Gait, M.J. 2003. Peptide-mediated cellular delivery of antisense oligonucleotides and their analogues. Cell. Mol. Life Sci. **60:** 1–10.
51. Zatsepin, T.S. *et al.* 2005. Conjugates of oligonucleotides and analogues with cell penetrating peptides as gene silencing agents. Curr. Pharm. Design **11:** 3639–3654.
52. Zorko, M. & U. Langel. 2005. Cell-penetrating peptides: mechanism and kinetics of cargo delivery. Adv. Drug Deliv. Rev. **57:** 529–545.
53. Astriab-Fisher, A. *et al.* 2002. Conjugates of antisense oligonucleotides with the Tat and Antennapedia cell-penetrating peptides: effect on cellular uptake, binding to target sequences, and biologic actions. Pharm. Res. **19:** 744–754.
54. Turner, J.J., A.A. Arzumanov & M.J. Gait. 2005. Synthesis, cellular uptake and HIV-1 Tat-dependent trans-activation inhibition activity of oligonucleotide analogues disulphide-conjugated to cell-penetrating peptides. Nucleic Acids Res. **33:** 27–42.
55. Turner, J.J. *et al.* 2005. Cell-penetrating peptide conjugates of peptide nucleic acids (PNA) as inhibitors of HIV-1 Tat-dependent trans-activation in cells. Nucleic Acids Res. **33:** 6837–6849.
56. Kaushik, N. *et al.* 2002. Anti-TAR polyamide nucleotide analog conjugated with a membrane-permeating peptide inhibits Human Immunodeficiency Virus Type I production. J. Virol. **76:** 3881–3891.
57. Chaubey, B. *et al.* 2005. A PNA-Transportan conjugate targeted to the TAR region of the HIV-1 genome exhibits both antiviral and virucidal properties. Virology **331:** 418–428.
58. Tripathi, S. *et al.* 2005. Anti-HIV-1 activity of anti-TAR polyamide nucleic acid conjugated with various membrane transducing peptides. Nucleic Acids Res. **33:** 4345–4356.
59. Boffa, L.C. *et al.* 2005. Therapeutically promising PNA complementary to a regulatory sequence for c-myc: pharmacokinetics in an animal model of human Burkitt's lymphoma. Oligonucleotides **15:** 85–93.
60. Soutschek, J., A. Akinc & B. Bramlage. 2004. Therapeutic silencing of an endogenous gene by systemic administration of modified siRNAs. Nature **432:** 173–178.
61. Asai, A. *et al.* 2003. A novel telomerase template antagonist (GRN163) as a potential anticancer agent. Cancer Res. **63:** 3931–3939.

Selection of Thioaptamers for Diagnostics and Therapeutics

XIANBIN YANG,[a] HE WANG,[b]
DAVID W. C. BEASLEY,[c] DAVID E. VOLK,[a] XU ZHAO,[a]
BRUCE A. LUXON,[a] LEE O. LOMAS,[b] NORBERT K. HERZOG,[c]
JUDITH F. ARONSON,[c] ALAN D. T. BARRETT,[c] JAMES F. LEARY,[d]
AND DAVID G. GORENSTEIN[a]

[a]*Department of Biochemistry and Molecular Biology, Sealy Center for Structural Biology, University of Texas Medical Branch at Galveston, Texas 77555, USA*

[b]*Ciphergen Biosystems, Fremont, California 94555, USA*

[c]*Department of Pathology, University of Texas Medical Branch at Galveston, Texas 77555, USA*

[d]*Basic Medical Sciences and Biomedical Engineering, Purdue University, West Lafayette, Indiana 47907, USA*

ABSTRACT: Thioaptamers offer advantages over normal phosphate ester backbone aptamers due to their enhanced affinity, specificity, and higher stability, largely due to the properties of the sulfur backbone modifications. Over the past several years, *in vitro* thioaptamer selection and bead-based thioaptamer selection techniques have been developed in our laboratory. Furthermore, several thioaptamers targeting specific proteins such as transcription factor NF-κB and AP-1 proteins have been identified. Selected thioaptamers have been shown diagnostic promise in proteome screens. Moreover, some promising thioaptamers have been shown in preliminary animal therapeutic dosing to increase survival in animal models of infection with West Nile virus.

KEYWORDS: aptamer; thioaptamer; oligonucleotide phosphorothioate; oligonucleotide phosphorodithioate; diagnostics; therapeutics; antiviral agent

RESULTS AND DISCUSSION

The utility of aptamers is reflected in the growing number of applications, which encompass the development of diagnostic reagents and therapeutic

Address for correspondence: David G. Gorenstein, 301 University Boulevard, Sealy Center for Structural Biology, University of Texas Medical Branch, Galveston, TX 77555–1157. Voice: 409-747-6800; fax: 409-747-6850.
e-mail: david@nmr.utmb.edu

Ann. N.Y. Acad. Sci. 1082: 116–119 (2006).
doi: 10.1196/annals.1348.065

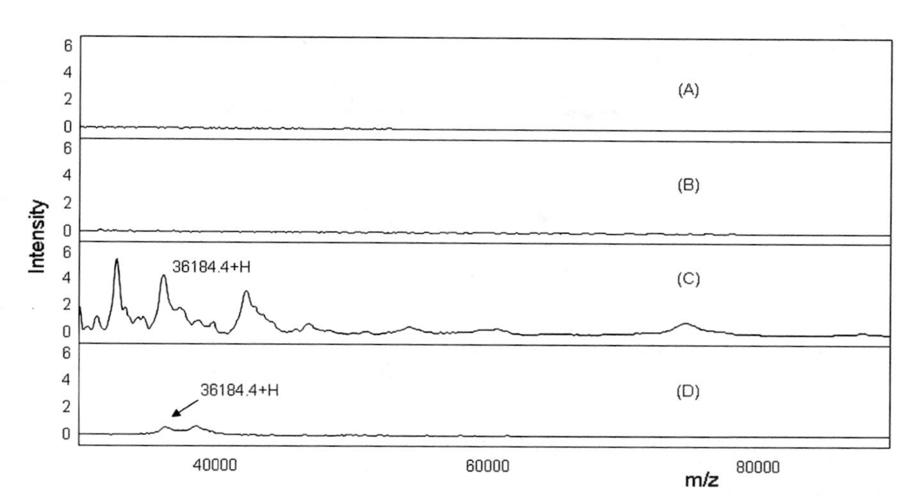

FIGURE 1. Mass spectra of captured proteins in the mass range of 30,000–90,000 m/z from crude stimulated nuclear extracts 70Z/3 when they were captured on different thioaptamer precoupled PU (reactive polyurethane) array surfaces. (**A**) no thioaptamer; (**B**) poly(IC)$_7$; (**C**) XBY-S2 and (**D**) XBY-6.

leads.[1,2] Because normal phosphate backbone aptamers are built from nucleotides joined by phosphate ester bonds, one chief problem is their susceptibility to nuclease hydrolysis in biological fluids; another potential limitation is that normal phosphate backbone aptamers may not bind tightly enough to achieve their biological effects. To overcome these shortcomings, we developed an *in vitro* combinatorial thioaptamer selection[3,4] and later a bead-based combinatorial library method[5] and high-throughput screening methods by flow cytometry,[6] that can simultaneously select for sequence and optimize mixed phosphoromonothioate, phosphorodithioate, or phosphate hybrid backbones.[5] We have, therefore, selected a number of thioaptamers against many proteins including transcription factor proteins, such as NF-κB,[5] the RNase H domain of HIV RT, and viral envelope proteins as well.

To develop potential thioaptamer diagnostic reagents, one of the AP-1 thioaptamers, XBY-S2, has been anchored on ProteinChip Array surfaces from Ciphergen Biosystems (Fremont, CA) to screen and identify bound proteins by mass spectrometry (FIG. 1). Briefly, a 5'-amino-linked AP-1 thioaptamer XBY-S2 was covalently bound to a PU (reactive polyurethane) array surface, and a group of proteins with sizes between 35–40 kDa bound preferentially to XBY-S2 (FIG. 1). Binding was greatly reduced when the anchored oligonucleotide sequences differed from XBY-S2, such as poly(IC)$_7$ and XBY-6 (an NF-κB thioaptamer). This result was further confirmed by a double competition assay. Some of the bound proteins have been identified by analysis of the MS/MS data from trypsin "on chip" digestion.

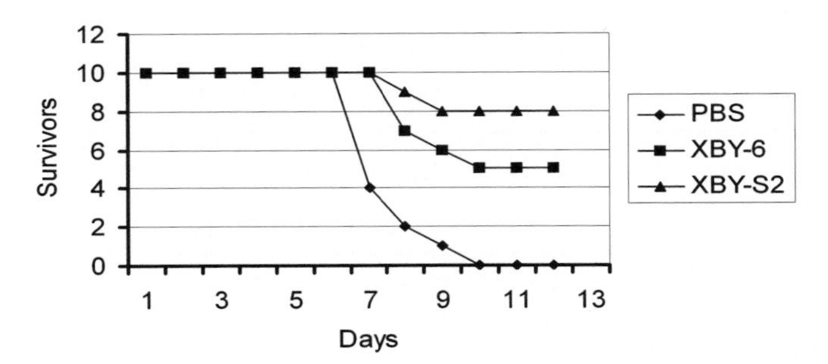

FIGURE 2. Survival rates of mice infected with West Nile virus and treated twice with PBS, XBY-6, or XBY-S2.

The thioaptamer XBY-S2 targeting AP-1 was also used to investigate antiviral activity against West Nile virus. As shown in FIGURE 2, control mice given phosphate-buffered saline solution (PBS) (or not shown, empty liposomes used to deliver the thioaptamer) succumbed to West Nile virus infection. However, 80% of mice treated with two liposomal doses of XBY-S2 (or 50% with an NF-κB thioaptamer, XBY-6) survived the challenge and remained healthy throughout the experiment, thus providing preliminary therapeutic dosing data demonstrating increased animal survival rates after infection with West Nile virus.

CONCLUSIONS

Efficient methods for identifying thioaptamers for their targets via *in vitro* thioaptamer selection or bead-based thioaptamer selection have been developed. Because of their potential for high affinity, high specificity, and high stability, the utility of thioaptamers in applications for diagnostics and therapeutics is encouraging.

ACKNOWLEDGMENTS

This research was supported by grants to Dr. D. Gorenstein from DARPA (P42296LS0000), NIH (U01 A1054827, IP30 ES06676), DoD (DADD13-02-C-007 and contract number W911 SR-04-C-0065 with the U.S. Edgewood Chemical Biological Center), the Welch Foundation (H-1296), and the State of Texas Advanced Technology Program (004952-0038-2003); to Dr. A. Barrett from NIH (U01 AI 60616); to Dr. J. Leary from NIH (R01 EB 00245); and to Dr. X. Yang from the Mike Hogg Fund. Two of the authors, David Gorenstein and Bruce Luxon have 75% equity interest in a startup company, AptaMed, which licensed technology described in the paper.

REFERENCES

1. BRODY, E.N. & L. GOLD. 2000. Aptamers as therapeutic and diagnostic agents. J. Biotechnol. **74:** 5–13.
2. YANG, X. & D.G. GORENSTEIN. 2004. Progress in thioaptamer development. Curr. Drug Targets **5:** 705–715.
3. KING, D.J. *et al.* 1998. Novel combinatorial selection of phosphorothioate oligonucleotide aptamers. Biochemistry **37:** 16489–16493.
4. BASSETT, S.E. *et al.* 2004. Combinatorial selection and edited combinatorial selection of phosphorothioate aptamers targeting human nuclear factor-kappaB RelA/p50 and RelA/RelA. Biochemistry **43:** 9105–9115.
5. YANG, X. *et al.* 2002. Construction and selection of bead-bound combinatorial oligonucleoside phosphorothioate and phosphorodithioate aptamer libraries designed for rapid PCR-based sequencing. Nucleic Acids Res. **30:** e132.
6. YANG, X. *et al.* 2003. Immunofluorescence assay and flow cytometric selection of bead bound aptamers. Nucleic Acids Res. **31:** e54.

Modification of the Pig CFTR Gene Mediated by Small Fragment Homologous Replacement

ROSALIE MAURISSE,[a] JUDY CHEUNG,[a] JONATHAN WIDDICOMBE,[b] AND DIETER C. GRUENERT[a,c,d]

[a]*California Pacific Medical Center Research Institute, San Francisco, California 94107, USA*

[b]*Departments of Human Physiology and Medicine, University of California, Davis, California 95616, USA*

[c]*Department of Medicine, University of Vermont, Burlington, Vermont 05405, USA*

[d]*Department of Laboratory Medicine, University of California, San Francisco, California 94107, USA*

ABSTRACT: The generation of a pig model of cystic fibrosis (CF) is a multistep process. Initial steps in this process involved the design and cloning of a small DNA fragment (SDF) or large oligodeoxynucleotide (LODN) that contains the F508del mutation and a silent restriction fragment length polymorphism causing mutation. This SDF/LODN was transfected into wild-type (WT) pig fetal fibroblast with the intention of modifying the pig genomic DNA by small fragment homologous replacement (SFHR). The targeted deletion (F508del) was detected in a subpopulation of transfected cells by allele-specific polymerase chain reaction (AS-PCR)

KEYWORDS: CFTR; gene modification; SFHR; site-directed mutagenesis

Sequence-specific modification of genomic DNA has been observed using various oligonucleotide-based gene targeting strategies[1,2] and can result in the restoration of gene function as well as guarantee that the corrected gene is expressed in "cell-appropriate" fashion. One strategy, small fragment homologous replacement (SFHR) has been shown to be able to modify specific regions of the cystic fibrosis (CF) transmembrane conductance regulator (CFTR)[3–7] gene. Both *in vitro* studies in human airway epithelial cells [3–7] and

Address for correspondence: Dieter C. Gruenert, Ph.D., California Pacific Medical Center Research Institute, 475 Brannan Street, Suite 220, San Francisco, CA 94107. Voice: 415-600-1362; fax: 415-600-1725.
e-mail: dieter@cpmcri.org

Ann. N.Y. Acad. Sci. 1082: 120–123 (2006). © 2006 New York Academy of Sciences.
doi: 10.1196/annals.1348.063

in vivo studies in mice[8] demonstrated that genomic CFTR could be modified by small DNA fragment/large oligodeoxynucleotide (SDF/LODN). These SFHR studies showed that it is possible to both correct the most common CFTR mutation, (F508del), as well as introduce it into the CFTR DNA of wild-type (WT) cells.

Given the potential of SFHR for sequence-specific modification of genomic CFTR and the recent advances in large animal cloning through somatic cell nuclear transfer (SCNT) into an oocyte,[9,10] it is now possible to consider the development of a large animal model of CF. The pig is an ideal candidate for CF, because it is both anatomically and physiologically similar to the human, most notably the airways.

The SDF was initially designed and cloned using a novel adaptation to the megaprimer site-directed mutagenesis strategy.[11–13] They were then transfected in WT pig fibroblasts. The transfected cells were analyzed to detect the modified cells. The mutations introduced into an SDF of the pig CFTR (pCFTR) gene, were a 3-bp deletion (del CTT) at codon 508 (the most common human CF mutation),[13] and a G>A transition mutation that gives rise to an SphI restriction enzyme cleavage site. The polymerase chain reaction (PCR) primers used to generate the mutant SDF were designed such that the desired mutation(s) was in the center of the primer. This design is particularly important to ensure annealing specificity. The first round PCR amplification of the genomic DNA from pig fetal fibroblast produced overlapping fragments A, B, and C (FIG. 1) and was carried out as three independent PCR amplifications with primers pairs indicated. The desired mutant SDF (SDF-F508del) was assembled in a second round of PCR by mixing 10 μL of each product A, B, and C from the first round of PCR. Sequence analysis confirmed the presence of the desired mutations. This straightforward method is faster and more efficient than previous PCR-based site-directed mutagenesis strategies.[14] It can simultaneously introduce at least two mutations into a DNA sequence using two rounds of PCR amplification with only a single gel purification step. Thus the approach described here is an improvement over previous methods[14] in that purification of the first round PCR product is not required.

Pig embryonic fibroblasts have been transfected with SDF-F508del (10^6 SDF/cell) using the Amaxa Nucleofector system (Gaithersburg, MD). The efficiency nucleofection transfection is 90% with very low cytotoxicity. Successive transfections were carried out every 2 days for a week; that is, at days 0, 2, and 4. DNA analysis was carried out at days 2 (after 1 cell transfection), 4 (after 2 cell transfections), and 9 (after 3 cell transfections).

Gel purified genomic DNA was analyzed by nested and allele-specific (AS)-PCRs. Control mixing experiments in which genomic DNA from WT cells and mutant SDF were combined indicated that a PCR artifact could be observed following AS-PCR, when $>10^4$ SDF/10^6 cell equivalents of DNA were mixed (no transfection) unless the genomic DNA is purified on an agarose

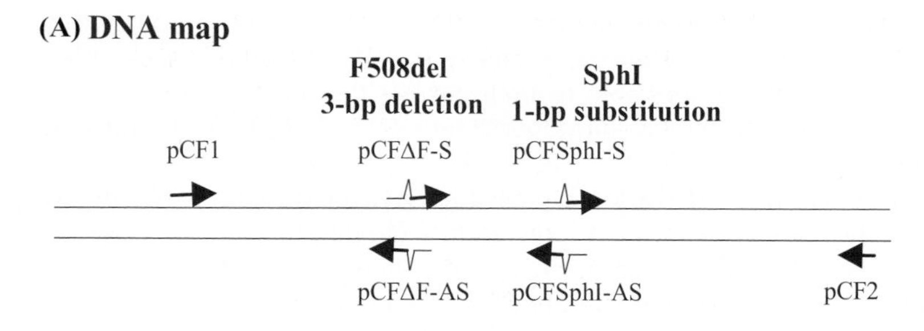

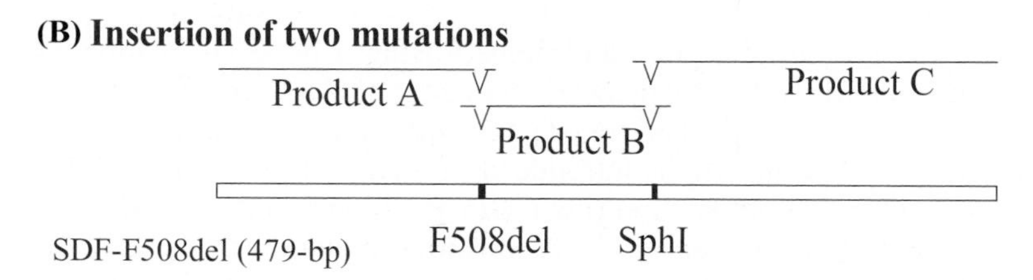

(C) DNA analysis AS-PCR

PCR product allele specific (443-bp)

pCF8Blna

pCF12

FIGURE 1. (**A**) Schematic representation of DNA primer localization with the arrows indicating the primer orientation. (**B**) In the first round of PCR, three independent PCR amplifications were carried out : product A synthesized with primers pair pCF1/pCFΔF-AS; product B with primers pairs pCFΔF-S/pCFSphI-AS, and product C with primers pairs pCFSphI-S/pCF2. PCR amplifications were used 100 ng of the template DNA in 100 μL reaction volume: 0.2 μM primers; 1X Buffer, 2.5 U high fidelity Accuprime Taq Polymerase (Invitrogen, Carlsbad, CA) in an MJ Research thermocycler (PCT-200, Waltham, MA). The amplification conditions were as follows: denaturation; 30 s at 95°C, annealing; 30 s at 55°C, elongation; 60 s at 68°C for 20 cycles with an 8-min elongation on the final cycle. Products A and B overlap by 27 nt at the 3′ and 5′ ends, respectively. Products B and C overlap by 20 nt at the 3′ and 5′ ends, respectively. In the second PCR, the SDF-F508del-SphI was generated by mixing 10 μl of products A, B, and C. This amplification introduces two mutations, the F508del (CTT deletion) and a G>A substitution that gives rise to a unique SphI restriction site. The fragments (A, B, and C) were added to a PCR mixture that contained 0.2 μM of the pCF2 and pCF1 primers, 1X PCR Buffer I, and 2.5 U Accuprime Taq high fidelity polymerase (Invitrogen). The PCR products were sized on a 1.5% agarose gel. The resultant 479-bp products (SDF-F508del) were gel purified (Qiagen, Germantown, MD) and cloned into a pCR-XL-topo vector (Invitrogen). (**C**) DNA analysis of WT cells transfected with SDF-F508del. AS-PCR was carried out using primers pCF8Blna specific to the mutated sequence (primer contains an lna base at the 3′ end) and pCF12 which is outside the SDF region.

gel. The transfected cells were analyzed by AS-PCR and nested PCR. Initial results indicate SFHR-mediated modification in these cells and suggest that a targeted sequence-specific deletion had occurred within a subpopulation of the transfected cells. Future studies will involve real-time PCR quantification of the gene modification efficiency and the clonal isolation of SFHR-modified cells.

REFERENCES

1. GRUENERT, D.C. *et al.* 2004. Oligonucleotide-based gene targeting approaches. Oligonucleotides **14:** 157–158; author reply 158–160.
2. PAREKH-OLMEDO, H. *et al.* 2005. Gene therapy progress and prospects: targeted gene repair. Gene Ther. **12:** 639–646.
3. KUNZELMANN, K. *et al.* 1996. Gene targeting of CFTR DNA in CF epithelial cells. Gene Ther. **3:** 859–867.
4. GONCZ, K.K. *et al.* 1998. Targeted replacement of normal and mutant CFTR sequences in human airway epithelial cells using DNA fragments. Hum. Mol. Genet. **7:** 1913–1919.
5. BRUSCIA, E. *et al.* 2002. Isolation of CF cell lines corrected at DeltaF508-CFTR locus by SFHR-mediated targeting. Gene Ther. **9:** 683–685.
6. GONCZ, K.K. *et al.* 2002. Application of SFHR to gene therapy of monogenic disorders. Gene Ther. **9:** 691–694.
7. COLOSIMO, A. *et al.* 2001. Targeted correction of a defective selectable marker gene in human epithelial cells by small DNA fragments. Mol. Ther. **3:** 178–185.
8. GONCZ, K.K. *et al.* 2001. Expression of DeltaF508 CFTR in normal mouse lung after site-specific modification of CFTR sequences by SFHR. Gene Ther. **8:** 961–965.
9. LAI, L. *et al.* 2002. Production of alpha-1,3-galactosyltransferase knockout pigs by nuclear transfer cloning. Science **295:** 1089–1092.
10. HOSHINO, Y. *et al.* 2005. Developmental competence of somatic cell nuclear transfer embryos reconstructed from oocytes matured in vitro with follicle shells in miniature pig. Cloning Stem Cells **7:** 17–26.
11. HEMSLEY, A. *et al.* 1989. A simple method for site-directed mutagenesis using the polymerase chain reaction. Nucleic Acids Res. **17:** 6545–6551.
12. XU, Z., A. COLOSIMO & D.C. GRUENERT. 2003. Site-directed mutagenesis using the megaprimer method. Methods Mol. Biol. **235:** 203–207.
13. KEREM, B. *et al.* 1989. Identification of the cystic fibrosis gene: genetic analysis. Science **245:** 1073–1080.
14. GE, L. & P. RUDOLPH. 1997. Simultaneous introduction of multiple mutations using overlap extension PCR. Biotechniques **22:** 28–30.

Nucleic Acid Therapeutics for Hematologic Malignancies—Theoretical Considerations

JOANNA B. OPALINSKA,[a] ANNA KALOTA,[b] JYOTI CHATTOPADHYAYA,[c] MASAD DAMHA,[d] AND ALAN M. GEWIRTZ[b]

[a]*Pommeranian Medical Academy, 71245 Szczecin, Poland*

[b]*Division of Hematology/Oncology, Department of Medicine and Abramson Family Cancer Research Institute, University of Pennsylvania School of Medicine, Philadelphia, Pennsylvania 19104, USA*

[c]*Department of Bioorganic Chemistry, Uppsala University, Biomedical Center, S-751 23 Uppsala, Sweden*

[d]*Department of Chemistry, McGill University, Montreal, QC, Canada H3A 2K6*

ABSTRACT: Our work is motivated by the belief that RNA targeted gene silencing agents can be developed into effective drugs for treating hematologic malignancies. In many experimental systems, antisense nucleic acids of various composition, including antisense oligodeoxynucleotides (AS ODNs) and short interfering RNA (siRNA), have been shown to perturb gene expression in a sequence specific manner. Nevertheless, our clinical experience, and those of others, have led us to conclude that the antisense nucleic acids (ASNAs) we, and others, employ need to be optimized with regard to intracellular delivery, targeting, chemical composition, and efficiency of mRNA destruction. We have hypothesized that addressing these critical issues will lead to the development of practical and effective nucleic acid drugs. An overview of our recent work which seeks to addresses these core issues is contained within this review.

KEYWORDS: leukemia; lymphoma; antisense; oligonucleotides

INTRODUCTION

While the advent of antibodies and small molecules has made an extraordinary difference in the lives of patients with chronic myelogenous leukemia (CML), and many lymphomas, patients with other hematologic malignancies have yet to enjoy the benefits of these types of targeted therapies, and the issue

Address for correspondence: Alan M. Gewirtz, M.D., Division of Hematology/Oncology, Department of Medicine and Abramson Family Cancer Research Institute, University of Pennsylvania School of Medicine, Philadelphia, PA 19104. Voice: 215-898-4499; fax: 215-573-2078.
e-mail: gewirtz@mail.med.upenn.edu

Ann. N.Y. Acad. Sci. 1082: 124–136 (2006).
doi: 10.1196/annals.1348.002

TABLE 1. Important issues to address in the development of improved RNA targeting drugs

1. Target biology—Physiologic role, expression level, half-life
2. Delivery
3. mRNA target—physical structure
4. Chemical composition of the molecule

of resistance to drugs like imatinib and rituximab is becoming increasingly important.[1–3] The ability to eliminate proteins that have heretofore escaped specific targeting, or have demonstated the ability to evolve resistant forms is the strength of an RNA-targeted, protein-eliminating, therapeutic strategy. In addition, an ever-expanding knowledge of the molecular pathogenesis of hematologic malignancies continues to suggest new therapeutic targets.[4] Finally, gene-silencing therapies are in principle highly specific, so that if the target gene is thoughtfully chosen, damage to nontargeted tissues should be minimized and a high therapeutic index should result.[5–7]

Numerous "gene-silencing" strategies have evolved over the years, and these have been primarily directed either to the genes themselves,[8–10] or to their messenger RNAs. Some exceptionally clever techniques for direct gene targeting have been developed[10,11] but they have not proven simple or reliable enough, at least thus far, for therapeutic applications. In contrast, the perceived ease with which mRNA can be targeted has resulted in most therapeutic efforts being directed to this approach.[12,13] A number of modalities are available for mRNA targeting and of these, the "antisense" strategies have been the most widely applied.[13–15] All are based on delivery of a reverse complementary, i.e., "antisense," nucleic acid strand into a cell expressing the gene of interest. By processes still unknown, the antisense nucleic acid (ASNA) strand and the mRNA target come into proximity and then hybridize if the strands are physically accessible to each other. Stable mRNA–ASNA duplexes can interfere with the splicing of heteronuclear RNA into mature messenger RNA,[16,17] block translation of mature mRNA,[18,19] or can lead to the destruction of the mRNA by the binding of endogenous nucleases, such as RNase H,[20,21] or by intrinsic enzymatic activity engineered into the sequence as is the case with ribozymes[22,23] and DNAzymes.[24,25] More recently, posttranscription gene silencing or RNA interference (RNAi)[26,27] has emerged as an exciting potential alternative to these more classical approaches.[14,15,28] However, it is quite clear that many of the therapeutic considerations that apply to the use of traditional antisense molecules will also apply to RNAi as evoked by short interfering RNA (siRNA), microRNA (miRNA), and short hairpin RNA expressed in viral vectors (shRNA). These issues are listed in TABLE 1 and include (1) choice of gene target; (2) development of rational, reliable targeting strategies; (3) stability of nucleic acid molecules in body fluids, and cells; and (4) ability to deliver improved molecules into cells.[29–34] The so-called "off target," or unintended, gene silencing is also being increasingly recognized with siRNA.[35–39]

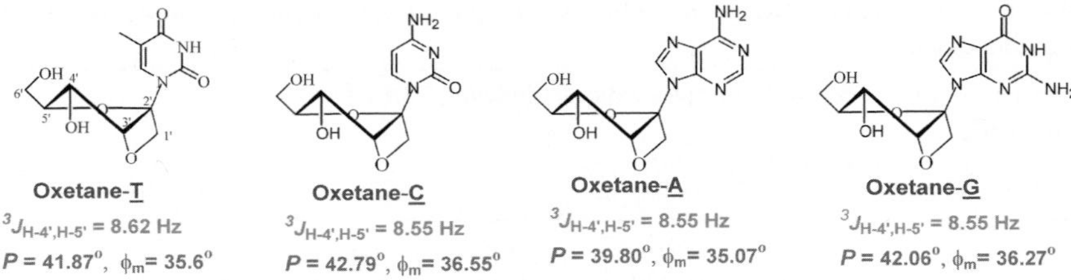

Oxetane-T
$^{3}J_{H\text{-}4',H\text{-}5'}$ = 8.62 Hz
P = 41.87°, ϕ_m= 35.6°

Oxetane-C
$^{3}J_{H\text{-}4',H\text{-}5'}$ = 8.55 Hz
P = 42.79°, ϕ_m= 36.55°

Oxetane-A
$^{3}J_{H\text{-}4',H\text{-}5'}$ = 8.55 Hz
P = 39.80°, ϕ_m= 35.07°

Oxetane-G
$^{3}J_{H\text{-}4',H\text{-}5'}$ = 8.55 Hz
P = 42.06°, ϕ_m= 36.27°

FIGURE 1. Structure of oxetane modified bases.

CHEMICAL MODIFICATIONS

Several promising new chemical modifications have been described recently.[40–44] Two relatively novel chemistries are of particular interest to us. One is the ribose sugar constraining oxetane modification[45,46](FIG. 1), the other is the 2′-deoxy-2′-fluoro-D-arabinonucleic acid (2′F-ANA) modification[47–49] (FIG. 2). The oxetane modification [oxetane, 1-(1′,3′-*O*-anhydro-ß-D-psicofuranosyl nucleosides)] imparts a number of highly desirable characteristics to oligodeoxynucleotides (ODN). These include enhanced nuclease stability and T_ms similar to those predicted for ODN/RNA hybrids. Since modification of all bases is not required to impart these characteristics, the ability of modified ODN to recruit RNase H is not impaired. We have examined the efficiency with which oxetane-modified antisense ODN inhibited *c-myb* gene expression in living cells and compared the results to standard phosphorothioate (PS)-modified ODN. We found that Myb mRNA and protein levels were equally diminished by oxetane and PS ODN, but the latter were delivered to cells with ~6× greater efficiency suggesting that oxetane-modified ODN were more potent on a molar basis.

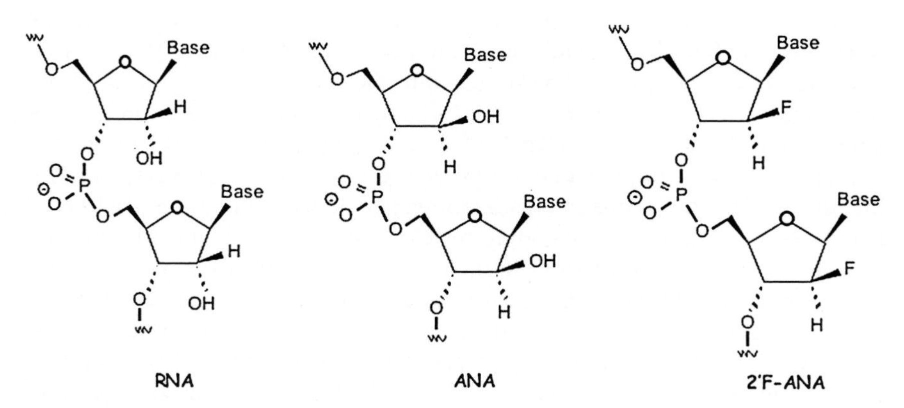

FIGURE 2. Derivation of 2′F-ANA.

The attraction of 2′F-ANA and 2′F-ANA-DNA chimeras derives from their nuclease resistance and their ability to simultaneously increase the strength of oligonucleotide:mRNA hybrids and elicit efficient RNase H-mediated degradation of target mRNA. In detailed studies we found that antisense oligonucleotides (AS ONs) containing a 2′F-ANA sugar modification not only performed as well as PS ODN, but had the added advantage of maintaining high intracellular concentrations for prolonged periods of time, which appears to promote longer-term gene silencing. To demonstrate this, we targeted the *c-myb* protooncogene's mRNA in human leukemia cells with fully PS 2′F-ANA-DNA chimeras (PS-2′FANA-DNA) and compared their gene-silencing efficiency with AS ON containing unmodified nucleosides (PS-DNA). When delivered by nucleofection, chemically modified ON of both types effected a greater than 90% knockdown of *c-myb* mRNA and protein expression, but the PS-2′F-ANA-DNA were able to accomplish this at 20% of the dose of PS-DNA, and in contrast to the PS-AS DNA, their silencing effect was still present 4 days after a single administration. This led us to conclude that PS-2′F-ANA-DNA chimeras are efficient gene-silencing molecules, and suggest that they could have significant therapeutic potential.

GENE TARGET SELECTION

With regard to gene target selection, our group has focused on short-lived mRNA molecules that encode proteins whose functions are critical to the targeted cell and that are equally short lived. We have hypothesized that such mRNA and protein targets are most efficiently eliminated from cells using the present methodologies, and importantly, are likely to have a considerable biologic impact on the cell that is being targeted. Examples of such targets include proteins encoded by the *c-myb* transcription factor gene, the transcriptional repressor *BCL-6*, the src family kinase lyn, and the *c-kit* receptor.[50–54] Of these, we have focused most of our efforts on *c-myb*, and an antisense DNA molecule targeting this transcription factor has been employed in the clinic for *ex vivo* bone marrow purging,[55] for systemic infusion into refractory leukemia patients, and will shortly be entering the clinic again in Phase I studies involving patients with a variety of hematologic malignancies.

Our experience with targeting the obligate hematopoietic transcription factor *c-myb*, whose mRNA and protein have very short half-lives of ~30–60 min each,[56] and which is required for G1/S transition, as well activation of other critical cellular genes,[53,55,57–61] would seem to be a good example.

To complement this strategy, we have also examined the utility of targeting an upstream signaling protein and a transcription factor simultaneously. In a specific instance, using cell lines and primary CML cells, we found that targeting *c-myb* and proto-Vav signaling protein can give additive cell inhibitory effects.[50]

mRNA MAPPING

How to select sequence targets within the mRNA of the candidate genes described above has been highly problematic. The physical structure of mRNAs is known to play an important role in target accessibility for both classical AS ONs (reviewed in Ref. 62), and more recently for siRNA molecules as well.[31] We have developed a novel approach to solving this problem, which depends on the use of fluorescent self-quenching reporter molecules (SQRM) to probe mRNA and signal the presence of hybridization accessible regions[51,63,64] (FIG. 3). We have used this method to target classical,[63] as well as chemically modified oligonucleotides.[42,51]

SQRM are DNA stem loops with fluorophore on the 5′ end (EDANS) and a quenching molecule (DABCYL) on the 3′ end.[51,63] To make the search more rational, and to ensure that strong hybrids were formed, we wrote a

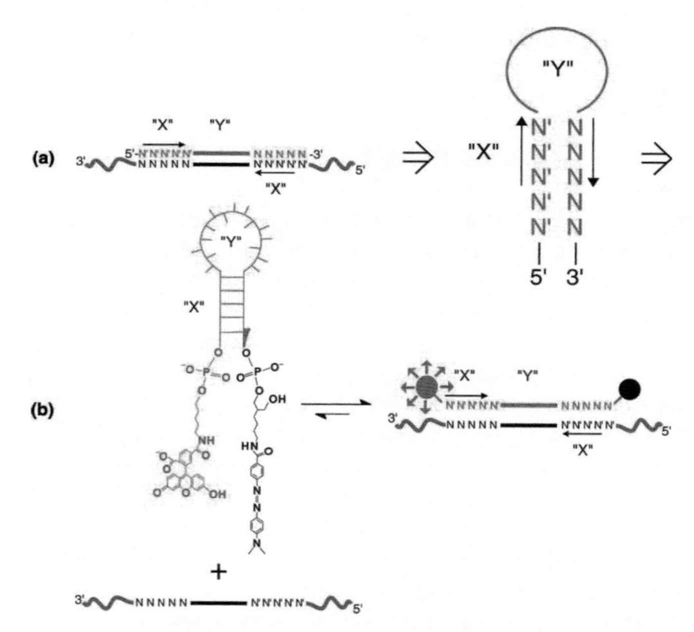

FIGURE 3. SQRM design and reaction. (**A**) Concept: to exploit the traditional stem–loop structure of the SQRMs, a computer algorithm ('AccessSearch') searches an entire sequence of mRNA for complementary sequences of a desired length (stems) that are separated by a proscribed distance (loop). (**B**) Chemistry: the complementary sequences are synthesized as SQRM possessing 50-fluorescein and 30 DABCYL groups. In the absence of target, quenching of fluorescence occurs. Once hybridization of the loop sequence to a complementary target takes place, the moieties are separated and fluorescence can be detected. From Gifford *et al.,* 2005. Nucleic Acids Res. 33: No. 3 e28. (Shown in color in online version).

simple computer algorithm that allowed us to search for inverted repeats within mRNA sequences downloaded from the NCBI site. Repeats of between 4 and 6 nucleotides were specified, along with intervening sequence of ~18–20 bases. Numerous sites compatible with these criteria could be found in any message we examined. SQRM corresponding to several such sites were synthesized, and then tested for hybridization as reported.[51] Of the probes studied, only two (+321, +964) demonstrated significant hybridization above background. Based on these results we would predict that a 26nt oligonucleotide targeted to *c-myb* beginning at +321 would effectively target the mRNA and silence gene expression. This prediction was evaluated in tissue culture and found to be correct (FIG. 4 A, B).

In another example, we have been developing antisense ODN and siRNA to knock down *BCL-6* expression in Diffuse Large B cell Lymphomas (DLCL). *BCL-6* is a zinc finger protein, which acts as a sequence-specific transcriptional repressor. Although *BCL-6* mRNA is ubiquitous, its expression is highest in germinal center B cells where it is thought to repress the expression of genes involved in B cell activation, cell cycle progression, and terminal differentiation.

In non-Hodgkin lymphomas, *BCL-6* is the most frequently deregulated gene and abnormal expression is found in ~30–40% of DLCL, and ~14% of follicular lymphomas (FL). Accordingly, we sought to develop gene-silencing antisense molecules targeted to *BCL-6* mRNA. Our strategy was based on the use of SQRM to rationally probe for hybridization accessible regions within a specific mRNA species. We found an accessible sequence within the *BCL-6* mRNA (SQRM-1310). An AS ON corresponding to SQRM-1310,

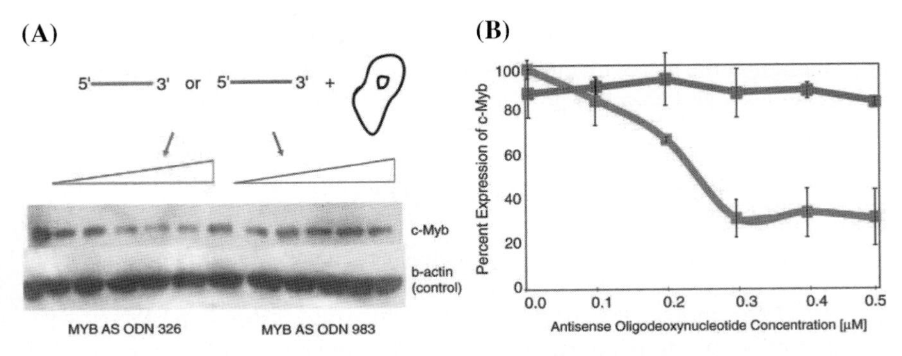

FIGURE 4. *c-myb* AS ODNs *in vivo*. An AS ODN corresponding to the SQRM321 was synthesized and transfected into hamster fibroblast Tkts 13 cells engineered to express human *c-myb*. (**A**) The Western blot shows a decrease in protein expression following treatment with the AS ODN 326–345 as compared with AS ODN 983–1000 (negative control). (**B**) Graphical representation of the western blot data: AS ODN 326 345 (dark gray); AS ODN 983–1000 (light gray). From Gifford *et al.*, 2005. Nucleic Acids Res. 2005, **33:** No. 3 e28. (Shown in color in online version).

were transfected into *BCL-6* (+) Louckes Cells using an Amaxa nucleoporator (Gaithersburg, MD). As a control, five other molecules were also transfected into Louckes Cells. Effects of these on *BCL-6* mRNA is shown in FIGURE 5. Cell viability was determined for 4 consecutive days. We found that cells transfected with Sequence +1310 exhibited an ~50% drop in viability within 24 h, while three other sequences were largely ineffective. Coincident with the viability drop, we found a seven-fold decrease in *BCL-6* mRNA in cells transfected with 1310, and little change in cells transfected with control ON. Corroborating Western Blot data on *BCL-6* expression were also obtained (FIG. 5).

DELIVERY

It is straightforward that without the ability to deliver material into cells, even the most cleverly designed molecule cannot be effective. As a general rule, oligonucleotides are taken up primarily through a combination of adsorptive and fluid-phase endocytosis.[65] After internalization, confocal and electron microscopy studies have indicated that the bulk of the oligonucleotides enter the endosome/lysosome compartment where most of the material either becomes trapped or degraded. Biologic inactivity is the predictable result of these events. Nevertheless, oligonucleotides can escape from the vesicles intact, enter the cytoplasm, and then diffuse into the nucleus where they presumably acquire their mRNA target. Colocalization of the effector strand and target mRNA in nucleoli,[23] or cytoplasmic P-bodies[66] appears important for AS ON and siRNA, respectively. In our hands and those of others,[67] lipid-transfecting agents have proven toxic to hematopoietic cells. Accordingly, we have begun to develop alternate means for delivering AS ON and siRNAs to hematopoietic cells including the use of electroporation for *ex vivo* delivery[52] (FIG. 6), as well as virosomes[68] and chitosan polymers[69,70] for systemic delivery.

CONCLUSIONS

The use of reverse complementary nucleic acid drugs to inhibit gene expression originated from studies that were initiated almost a quarter of a century ago.[71,72] Even though the mechanism by which these drugs modulate gene expression is not always clear,[73–75] the clinical development of nucleic acid drugs has proceeded to the point at which several of these drugs have entered Phase I/II, and in a few cases, Phase III trials. Results to date for most of these trials have been disappointing from the point of view of clinical efficacy. Nonetheless, the attraction for drugs of this class remains very strong and has been revitalized the development of RNAi.[76] This very exciting approach to gene silencing is, at the end of the day, also and "antisense-based

(A)

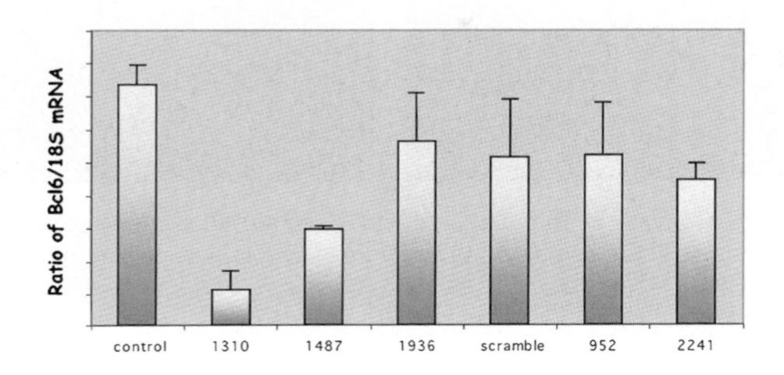

(B)

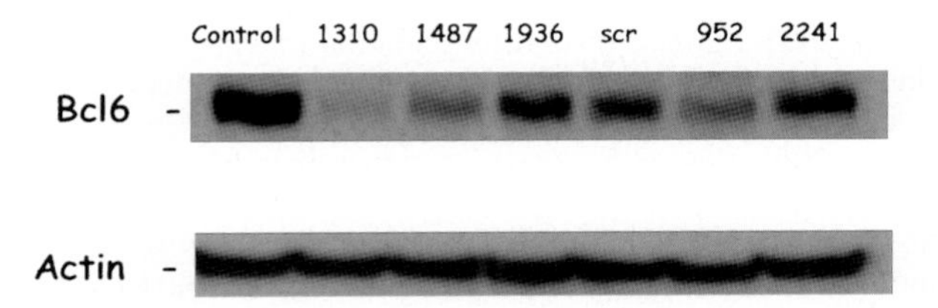

FIGURE 5. BCL-6 mRNA (**A**) and protein levels (**B**) in Louckes Cells treated with ODN targeted to different sequences within the *BCL-6* mRNA.

methodology" whose robustness in a clinical setting also needs to be determined. Therefore, despite the fact that only one ASNA drug has received FDA approval to date,[77] there is reason to remain optimistic that the problems that slow down progress in this field will be overcome, and that many very useful drugs for the treatment of a variety of human and animal diseases will result.

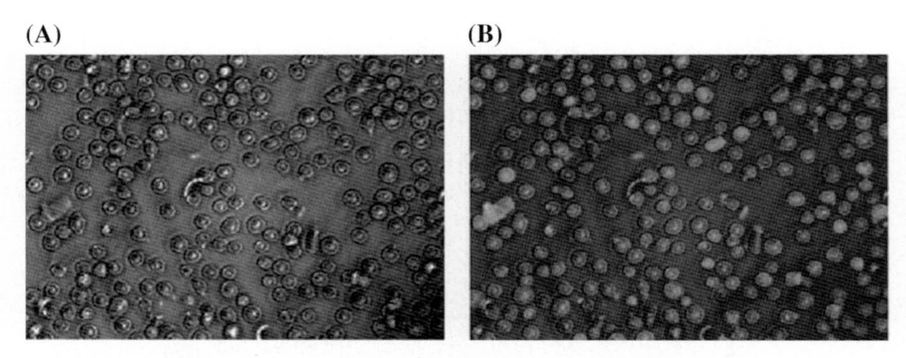

FIGURE 6. Delivery of fluorescein labeled unmodified ODNs into K562 cells with the nucleoporation technique: phase (**A**) and fluorescent (**B**) low power (200x) photomicrographs. From Opalinska *et al.*, 2004. Nucleic Acids Res. **32:** 5791-5799. (Shown in color in online version).

ACKNOWLEDGMENTS

This study was supported in part by grants from the National Cancer Institute, and the Doris Duke Charitable Foundation.

REFERENCES

1. TERUI, Y. *et al.* 1995. Expression of differentiation-related phenotypes and apoptosis are independently regulated during myeloid cell differentiation. J. Biochem. (Tokyo) **117:** 77–84.
2. SAWYERS, C.L. 2001. Research on resistance to cancer drug Gleevec. Science **294:** 1834.
3. FRIEDBERG, J.W. 2005. Unique toxicities and resistance mechanisms associated with monoclonal antibody therapy. Hematology (Am. Soc. Hematol. Educ. Program), 329–334.
4. FROHLING, S. *et al.* 2005. Genetics of myeloid malignancies: pathogenetic and clinical implications. J. Clin. Oncol. **23:** 6285–6295.
5. HERMISTON, T. 2000. Gene delivery from replication-selective viruses: arming guided missiles in the war against cancer. J. Clin. Invest. **105:** 1169–1172.
6. NETTELBECK, D.M., V. JEROME & R. MULLER. 2000. Gene therapy: designer promoters for tumour targeting. Trends Genet. **16:** 174–181.
7. VILE, R.G., S.J. RUSSELL & N.R. LEMOINE. 2000. Cancer gene therapy: hard lessons and new courses. Gene Ther. **7:** 2–8.
8. CASEY, B.P. & P.M. GLAZER. 2001. Gene targeting via triple-helix formation. Prog. Nucleic Acid Res. Mol. Biol. **67:** 163–192.
9. URBACH, A.R. & P.B. DERVAN. 2001. Toward rules for 1:1 polyamide:DNA recognition. Proc. Natl. Acad. Sci. USA **98:** 4343–4348.
10. DURAI, S. *et al.* 2005. Zinc finger nucleases: custom-designed molecular scissors for genome engineering of plant and mammalian cells. Nucleic Acids Res. **33:** 5978–5990.

11. DERVAN, P.B. & R.W. BURLI. 1999. Sequence-specific DNA recognition by polyamides. Curr. Opin. Chem. Biol. **3:** 688–693.
12. GEWIRTZ, A.M. 1998. Antisense oligonucleotide therapeutics for human leukemia. Curr. Opin. Hematol. **5:** 59–71.
13. OPALINSKA, J.B. & A.M. GEWIRTZ. 2002. Nucleic-acid therapeutics: basic principles and recent applications. Nat. Rev. Drug Discov. **1:** 503–514.
14. HEXNER, E.O. & A.M. GEWIRTZ. 2005. RNA interference for treating haematological malignancies. Expert Opin. Biol. Ther. **5:** 1585–1592.
15. ROSSI, J.J. 2005. RNAi therapeutics: SNALPing siRNAs *in vivo*. Gene Ther. **13:** 583–584.
16. KOLE, R. & P. SAZANI. 2001. Antisense effects in the cell nucleus: modification of splicing. Curr. Opin. Mol. Ther. **3:** 229–234.
17. DOMINSKI, Z. & R. KOLE. 1994. Identification and characterization by antisense oligonucleotides of exon and intron sequences required for splicing. Mol. Cell Biol. **14:** 7445–7454.
18. SUMMERTON, J. *et al*. 1997. Morpholino and phosphorothioate antisense oligomers compared in cell-free and in-cell systems. Antisense Nucleic Acid Drug Dev. **7:** 63–70.
19. IVERSEN, P.L. 2001. Phosphorodiamidate morpholino oligomers: favorable properties for sequence-specific gene inactivation. Curr. Opin. Mol. Ther. **3:** 235–238.
20. ZAMARATSKI, E., P.I. PRADEEPKUMAR & J. CHATTOPADHYAYA. 2001. A critical survey of the structure-function of the antisense oligo/RNA heteroduplex as substrate for RNase H. J. Biochem. Biophys. Methods **48:** 189–208.
21. CROOKE, S.T. 1998. Molecular mechanisms of antisense drugs: RNase H. Antisense Nucleic Acid Drug Dev. **8:** 133–134.
22. CASTANOTTO, D., M. SCHERR & J.J. ROSSI. 2000. Intracellular expression and function of antisense catalytic RNAs. Methods Enzymol. **313:** 401–420.
23. ROSSI, J.J. 1999. Ribozymes in the nucleolus. Science **285:** 1685.
24. SANTORO, S.W. & G.F. JOYCE. 1997. A general purpose RNA-cleaving DNA enzyme. Proc. Natl. Acad. Sci. USA **94:** 4262–4266.
25. WU, Y. *et al*. 1999. Inhibition of bcr-abl oncogene expression by novel deoxyribozymes (DNAzymes). Hum. Gene Ther. **10:** 2847–2857.
26. HANNON, G.J. 2002. RNA interference. Nature **418:** 244–251.
27. MARTINEZ, J. *et al*. 2002. Single-stranded antisense siRNAs guide target RNA cleavage in RNAi. Cell **110:** 563–574.
28. GRUNWELLER, A. & R.K. HARTMANN. 2005. RNA interference as a gene-specific approach for molecular medicine. Curr. Med. Chem. **12:** 3143–3161.
29. CEJKA, D., D. LOSERT & V. WACHECK. 2006. Short interfering RNA (siRNA): tool or therapeutic? Clin. Sci. (Lond.) **110:** 47–58.
30. LIAO, H. & J.H. WANG. 2005. Biomembrane-permeable and ribonuclease-resistant siRNA with enhanced activity. Oligonucleotides **15:** 196–205.
31. BROWN, K.M., C.Y. CHU & T.M. RANA. 2005. Target accessibility dictates the potency of human RISC. Nat. Struct. Mol. Biol. **12:** 469–470.
32. PATZEL, V. *et al*. 2005. Design of siRNAs producing unstructured guide-RNAs results in improved RNA interference efficiency. Nat. Biotechnol. **23:** 1440–1444.
33. SPAGNOU, S., A.D. MILLER & M. KELLER. 2004. Lipidic carriers of siRNA: differences in the formulation, cellular uptake, and delivery with plasmid DNA. Biochemistry **43:** 13348–13356.

34. Sioud, M. 2005. siRNA delivery *in vivo*. Methods Mol. Biol. **309:** 237–249.
35. Lassus, P., J. Rodriguez & Y. Lazebnik. 2002. Confirming specificity of RNAi in mammalian cells. Sci. STKE **2002:** PL13.
36. Hornung, V. *et al.* 2005. Sequence-specific potent induction of IFN-alpha by short interfering RNA in plasmacytoid dendritic cells through TLR7. Nat. Med. **11:** 263–270.
37. Zhang, Y.C. *et al.* 2005. Antisense inhibition: oligonucleotides, ribozymes, and siRNAs. Methods Mol. Med. **106:** 11–34.
38. Bitko, V. *et al.* 2005. Inhibition of respiratory viruses by nasally administered siRNA. Nat. Med. **11:** 50–55.
39. Zhang, Z. *et al.* 2005. siRNA binding proteins of microglial cells: PKR is an unanticipated ligand. J. Cell Biochem. **97:** 1217–1229.
40. Grunweller, A. *et al.* 2003. Comparison of different antisense strategies in mammalian cells using locked nucleic acids, 2′-O-methyl RNA, phosphorothioates and small interfering RNA. Nucleic Acids Res. **31:** 3185–3193.
41. Jepsen, J.S., M.D. Sorensen & J. Wengel. 2004. Locked nucleic acid: a potent nucleic acid analog in therapeutics and biotechnology. Oligonucleotides **14:** 130–146.
42. Kalota, A. *et al.* 2006. 2′-deoxy-2′-fluoro-beta-D-arabinonucleic acid (2′F-ANA) modified oligonucleotides (ON) effect highly efficient, and persistent, gene silencing. Nucleic Acids Res. **34:** 451–461.
43. Kang, H. *et al.* 2004. Inhibition of MDR1 gene expression by chimeric HNA antisense oligonucleotides. Nucleic Acids Res. **32:** 4411–4419.
44. Prakash, T.P. *et al.* 2004. 2′-O-[2-[(N,N-dimethylamino)oxy]ethyl]-modified oligonucleotides inhibit expression of mRNA *in vitro* and *in vivo*. Nucleic Acids Res. **32:** 828–833.
45. Opalinska, J.B. *et al.* 2004. Oxetane modified, conformationally constrained, antisense oligodeoxyribonucleotides function efficiently as gene silencing molecules. Nucleic Acids Res. **32:** 5791–5799.
46. Pradeepkumar, P.I. & J. Chattopadhyaya. 2001. Oxetane modified antisense oligonucleotides promote RNase H cleavage of the complementary RNA strand in the hybrid duplex as efficiently as the native, and offer improved endonuclease resistance. JCS. Perkin **2:** 2074–2083.
47. Wilds, C.J. & M.J. Damha. 2000. 2′-Deoxy-2′-fluoro-beta-D-arabinonucleosides and oligonucleotides (2′F-ANA): synthesis and physicochemical studies. Nucleic Acids Res. **28:** 3625–3635.
48. Viazovkina, E., M.M. Mangos & M.J. Damha. 2003. Synthesis and physicochemical properties of 2′-deoxy-2′,2″-difluoro-beta-D-ribofuranosyl and 2′-deoxy-2′,2″-difluoro-alpha-D-ribofuranosyl oligonucleotides. Nucleosides Nucleotides Nucleic Acids **22:** 1251–1254.
49. Mangos, M.M. *et al.* 2003. Efficient RNase H-directed cleavage of RNA promoted by antisense DNA or 2′F-ANA constructs containing acyclic nucleotide inserts. J. Am. Chem. Soc. **125:** 654–661.
50. Opalinska, J.B. *et al.* 2005. Multigene targeting with antisense oligodeoxynucleotides: an exploratory study using primary human leukemia cells. Clin. Cancer Res. **11:** 4948–4954.
51. Gifford, L.K. *et al.* 2005. Identification of antisense nucleic acid hybridization sites in mRNA molecules with self-quenching fluorescent reporter molecules. Nucleic Acids Res. **33:** e28.
52. Ptasznik, A. *et al.* 2004. Short interfering RNA (siRNA) targeting the Lyn kinase

induces apoptosis in primary, and drug-resistant, BCR-ABL1(+) leukemia cells. Nat. Med. **10:** 1187–1189.
53. GEWIRTZ, A.M. 1999. Myb targeted therapeutics for the treatment of human malignancies. Oncogene **18:** 3056–3062.
54. RATAJCZAK, M.Z. *et al.* 1992. Role of the KIT protooncogene in normal and malignant human hematopoiesis. Proc. Natl. Acad. Sci. USA **89:** 1710–1714.
55. LUGER, S.M. *et al.* 2002. Oligodeoxynucleotide-mediated inhibition of c-myb gene expression in autografted bone marrow: a pilot study. Blood **99:** 1150–1158.
56. BIES, J., V. NAZAROV & L. WOLFF. 1999. Identification of protein instability determinants in the carboxy-terminal region of c-Myb removed as a result of retroviral integration in murine monocytic leukemias. J. Virol. **73:** 2038–2044.
57. GEWIRTZ, A.M. & B. CALABRETTA. 1988. A c-myb antisense oligodeoxynucleotide inhibits normal human hematopoiesis *in vitro*. Science **242:** 1303–1306.
58. GEWIRTZ, A.M. *et al.* 1989. G1/S transition in normal human T-lymphocytes requires the nuclear protein encoded by c-myb. Science **245:** 180–183.
59. NESS, S.A., A. MARKNELL & T. GRAF. 1989. The v-myb oncogene product binds to and activates the promyelocyte-specific mim-1 gene. Cell **59:** 1115–1125.
60. CALABRETTA, B. & A.M. GEWIRTZ. 1991. Functional requirements of c-myb during normal and leukemic hematopoiesis. Crit. Rev. Oncog. **2:** 187–194.
61. SHETZLINE, S.E. *et al.* 2004. Neuromedin U: a Myb-regulated autocrine growth factor for human myeloid leukemias. Blood **104:** 1833–1840.
62. GEWIRTZ, A.M., D.L. SOKOL & M.Z. RATAJCZAK. 1998. Nucleic acid therapeutics: state of the art and future prospects. Blood **92:** 712–736.
63. SOKOL, D.L. *et al.* 1998. Real time detection of DNA. RNA hybridization in living cells. Proc. Natl. Acad. Sci. USA **95:** 11538–11543.
64. OPALINSKA, J.B. & A.M. GEWIRTZ. 2005. Rationally targeted, conformationally constrained, oxetane-modified oligonucleotides demonstrate efficient gene-silencing activity in a cellular system. Ann. N. Y. Acad. Sci. **1058:** 39–51.
65. BELTINGER, C. *et al.* 1995. Binding, uptake, and intracellular trafficking of phosphorothioate-modified oligodeoxynucleotides. J. Clin. Invest. **95:** 1814–1823.
66. ROSSI, J.J. 2005. RNAi and the P-body connection. Nat. Cell Biol. **7:** 643–644.
67. FENSKE, D.B. & P.R. CULLIS. 2005. Entrapment of small molecules and nucleic acid-based drugs in liposomes. Methods Enzymol. **391:** 7–40.
68. SCHOEN, P. *et al.* 1999. Gene transfer mediated by fusion protein hemagglutinin reconstituted in cationic lipid vesicles. Gene Ther. **6:** 823–832.
69. KOPING-HOGGARD, M. *et al.* 2001. Chitosan as a nonviral gene delivery system. Structure-property relationships and characteristics compared with polyethylenimine *in vitro* and after lung administration *in vivo*. Gene Ther. **8:** 1108–1121.
70. SINGLA, A.K. & M. CHAWLA. 2001. Chitosan: some pharmaceutical and biological aspects–an update. J. Pharm. Pharmacol. **53:** 1047–1067.
71. PATERSON, B.M. & J.O. BISHOP. 1977. Changes in the mRNA population of chick myoblasts during myogenesis *in vitro*. Cell **12:** 751–765.
72. STEPHENSON, M.L. & P.C. ZAMECNIK. 1978. Inhibition of Rous sarcoma viral RNA translation by a specific oligodeoxyribonucleotide. Proc. Natl. Acad. Sci. USA **75:** 285–288.
73. GEWIRTZ, A.M. 2000. Oligonucleotide therapeutics: a step forward. J. Clin. Oncol. **18:** 1809–1811.

74. SCHERR, M. *et al.* 2000. RNA accessibility prediction: a theoretical approach is consistent with experimental studies in cell extracts. Nucleic Acids Res. **28:** 2455–2461.
75. STEIN, C.A. 1995. Does antisense exist? Nat. Med. **1:** 1119–1121.
76. SHUEY, D.J., D.E. MCCALLUS & T. GIORDANO. 2002. RNAi: gene-silencing in therapeutic intervention. Drug Discov. Today **7:** 1040–1046.
77. DE SMET, M.D., C.J. MEENKEN & G.J. VAN DEN HORN. 1999. Fomivirsen – a phosphorothioate oligonucleotide for the treatment of CMV retinitis. Ocul. Immunol. Inflamm. **7:** 189–198.

CpG Oligonucleotides Improve the Protective Immune Response Induced by the Licensed Anthrax Vaccine

DENNIS M. KLINMAN,[a] HANG XIE,[a] AND BRUCE E. IVINS[b]

[a] *Section of Retroviral Immunology, Center for Biologics Evaluation and Research, Food and Drug Administration, Bethesda, Maryland 20892, USA*

[b] *Bacteriology Division, United States Army Medical Research Institute of Infectious Diseases, Fort Detrick, Frederick, Maryland 21702, USA*

ABSTRACT: Synthetic oligodeoxynucleotides (ODN) containing unmethylated CpG motifs act as immune adjuvants, improving the response elicited by a coadministered vaccine. Combining CpG ODN with anthrax vaccine adsorbed (AVA, the licensed human vaccine) increases the speed, magnitude, and avidity of the resultant antibody response. IgG Abs against anthrax protective antigen (PA) protect mice, guinuea pigs, and rhesus macaques from infection.

KEYWORDS: anthrax; vaccine; adjuvant; CpG oligonucleotide; protection

INTRODUCTION

Bacillus anthracis (*B. anthracis)* is an aerobic gram-positive spore-forming bacterium found naturally in wild and domesticated animals.[1] It is highly resistant to environmental degradation, and produces a tripartite toxin that reduces the ability of the host's immune system to eliminate the pathogen.[1] Human exposure to anthrax typically arises following contact with infected livestock, and generally results in a mild form of cutaneous disease.[1,2] However, anthrax spores designed for aerosol delivery were intentionally released in the United States by bioterrorists in 2001. The resultant morbidity, mortality, and widespread panic underscored the potential for anthrax to be used as a bioterror agent, and the need to improve the speed, magnitude, and safety of anthrax vaccination.[3]

Vaccination is the least costly and most effective method of reducing susceptibility to anthrax and of accelerating the development of protective immunity following pathogen exposure.[2] Anthrax vaccine adsorbed (AVA) is the only

Address for correspondence: Dennis Klinman, M.D., Ph.D., Bldg. 29A Rm. 3 D 10, CBER/FDA, Bethesda, MD 20892. Voice: 301-827-1707; fax: 301-496-1810.
e-mail: Klinman@CBER.FDA.GOV

Ann. N.Y. Acad. Sci. 1082: 137–150 (2006). © 2006 New York Academy of Sciences.
doi: 10.1196/annals.1348.030

anthrax vaccine licensed for human use in the United States. It is prepared by adsorbing the culture filtrate of an attenuated toxinogenic nonencapsulated strain of *B. anthracis* (V770-NP1-R) onto aluminum hydroxide.[4] The immunologically dominant component of AVA is "protective antigen" (PA), which is critical to the activity of anthrax toxin. Antibodies (Ab) against PA neutralize the toxin, inhibit spore germination, and improve the phagocytosis/killing of spores by macrophages.[5–8] Optimal vaccination with AVA involves a series of six immunizations delivered over 18 months followed by yearly boosters.[9,10] This schedule may adversely effect the host by causing joint pain, gastrointestinal disorders, and pneumonia, leading many U.S. soldiers to refuse vaccination.[9,11,12] Strategies that reduce the dose and/or number of AVA immunizations could improve the safety and compliance profile of this vaccine.

Synthetic oligodeoxynucleotides (ODN) containing immunostimulatory "CpG motifs" have been shown to act as vaccine adjuvants, improving the immune response to a variety of coadministered antigens.[13–15] CpG ODN interact with toll-like receptor 9 expressed by human B cells and plasmacytoid dendritic cells,[16–19] improving antigen presentation and triggering the production of Th1 and proinflammatory cytokines and chemokines (including interferon-γ [IFN-γ], interleukin-6 [IL-6], IL-12, IL-18, and tumor necrosis factor-α [TNF-α]).[16,17,20,21] Most studies examining the adjuvant activity of CpG ODN have been conducted in mice.[22–26] However, due to evolutionary divergence in CpG recognition between species, ODN that are highly active in rodents may be poorly immunostimulatory in primates.[27–29] Thus, preclinical studies examining whether CpG ODN can accelerate and boost the immune response elicited by AVA are best conducted in a relevant primate model.

In this context, several structurally distinct "classes" of CpG ODN have been identified that differentially stimulate primate immune cells.[29,30] "D" type ODN (which contain a single CpG motif linked to a 3' poly-G tail) trigger plasmacytoid dendritic cells (pDC) to produce high levels of IFN-α but have little effect on B cells.[29] In contrast, "K" type ODN (which express multiple CpG motifs but lack a poly-G tail) stimulate B cells to produce IgM and IL-6, and trigger pDC to produce TNF-α but not IFN-α.[29]

This article reviews the effect of coadministering "K" class CpG ODN with AVA on the rapidity, titer, affinity, and protective efficacy of the IgG anti-PA response elicited in rodents and primates.

MATERIALS AND METHODS

Reagents

Phosphorothioate ODNs were synthesized at the Center for Biologics Core Facility, and had no detectable protein or endotoxin contamination.[31] ODN 7909 was provided by Coley Pharmaceuticals (Wellesley, MA). AVA was obtained from BioPort Corporation (East Lansing, MI). Recombinant PA (rPA) and toxinogenic ($pXO1^{+}$) and nonencapsulated ($pXO2^{-}$) Sterne vaccine strain spores of *B. anthracis* (STI) were provided by USAMRIID (Fort Detrick, MD).

Animals

All animal studies were ACUC approved and were conducted in AAALAC accredited facilities. Animals were monitored daily by veterinarians. Specific pathogen-free A/J mice were obtained from the Jackson Laboratories (Bar Harbor, ME) and housed in sterile microisolator cages in a barrier environment.

Hartley guinea pigs, 325–375 g (Charles River, Wilmington, MA) were immunized i.m. with 0.5 mL doses of AVA plus 100–300 μg of an equimolar mixture of CpG ODNs and boosted with the same material 4 weeks later.

Healthy 3-year-old female rhesus macaques were obtained from the FDA colony in South Carolina. In the first primate experiment, animals (3–5 kg, n = 6/group) were immunized s.c. at 0 and 6 weeks with 0.5 mL of AVA plus 250 μg of an equimolar mixture of 3 CpG ODN (described below), and then "challenged" with 10^5 STI anthrax spores when serum anti-PA titers returned to base line at week 26. Serum Ab levels were monitored 3 weeks post immunization and 2 weeks post boost/challenge. In the second experiment, macaques (3–5 kg, n = 5/group) were immunized s.c. with 0.5 mL of AVA plus 500 μg of ODN 7909 or the same mixture of 3 CpG ODN mentioned above. Animals were bled at least weekly for 6 weeks.

Challenge Studies

A/J mice were immunized i.p. with AVA formulated in alum ± CpG ODN. The mice were bled weekly, and their serum was stored at –20° C until use. Mice were challenged i.p. with $3 \times 10^2 - 9 \times 10^4$ 50% lethal dose(s) (LD_{50}) of STI spores suspended in 0.5 mL sterile phosphate-buffered saline (PBS) (1 $LD_{50} = 1.1 \times 10^3$ STI spores). Survival was monitored for 21 days.

Guinea pigs were challenged i.m. 10 week post immunization with 5000 (50 LD_{50}) *B. anthracis* Ames spores.

For serum transfer studies, serum from all monkeys in each treatment group was pooled. One hundred microliters of pooled serum was injected i.p. into 6-week-old male A/J mice (n = 20/group). The following day, mice were challenged i.p. with 500 μL of PBS containing 30 LD_{50} STI anthrax spores. Mice were monitored daily for 2 weeks, and time to death was recorded.

Ab Assays

IgG anti-PA titers were measured by coating 96-well microtiter plates (Reacti-Bind EIA plates; Pierce Endogen, Rockford, IL) with 1 μg/mL of rPA. Serum samples were serially diluted in PBS plus 5% nonfat dry milk and incubated on these plates for 1 h at 37°C. The plates were washed, and for avidity assays overlaid for 15 min with 200 μL of 6 M urea. Bound Ab was detected using peroxidase-conjugated goat anti-human IgG (Kirkegaard & Perry, Gaithersburg, MD) followed by ABTS substrate (Kirkegaard & Perry). For avidity comparisons, titers were determined by comparison to a standard curve generated using high-titered anti-PA serum. Toxin-neutralizing titers were measured as previously described.[32] All samples were analyzed in triplicate.

Statistics

Differences in the kinetic development of anti-PA immune responses were determined by two-way ANOVA. Differences in the IgG anti-PA response induced by various vaccine-adjuvant combinations were assessed by one-way ANOVA. Differences in survival were evaluated using χ^2 analysis of Kaplan–Meier curves. Correlation coefficients were determined by linear regression analysis. The predictive value of IgG anti-PA and toxin-neutralizing titers on survival was evaluated using two-parameter logistic regression.[33]

RESULTS

CpG ODN Improve the Immunogenicity of AVA in Mice

A/J mice were immunized with increasing doses of AVA ± 20 μg of CpG ODN 1555 (a safe and biologically active dose of CpG ODN[22]). A/J mice were selected for study because they are susceptible to challenge by attenuated STI anthrax spores, allowing the protective activity of the resultant immune response to be examined in a BL-2 facility.[32] As the dose of AVA increased, animals co-immunized with CpG ODN rapidly generated significantly higher titers of IgG anti-PA Abs than mice immunized with AVA alone ($P < 0.01$, FIG. 1). Vaccinated mice were challenged 2 weeks later with approximately 100 LD_{50} of STI spores. A significantly greater fraction of mice immunized with AVA + CpG ODN survived challenge (29/39, 74%) than those immunized with AVA alone (12/46, 26%, $P < 0.01$) or AVA + control ODN (2/12, 17%, $P < 0.01$).

Humoral Immunity as a Predictor of Protection

There is a considerable interest in identifying a surrogate marker for protective immunity against anthrax. Toward that end, the toxin-neutralizing activity (TNA) and IgG anti-PA titer of serum from vaccinated mice were evaluated for

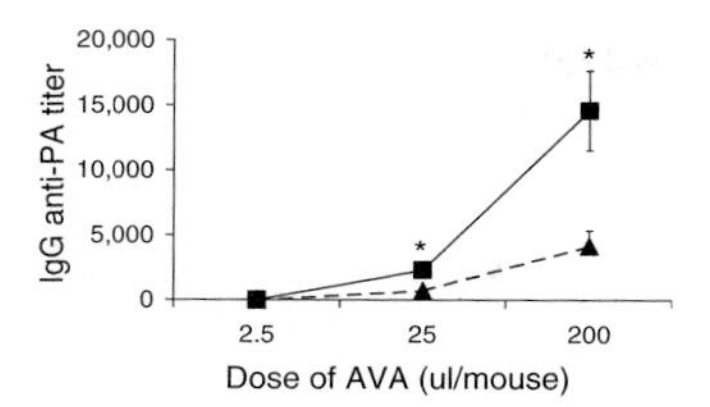

FIGURE 1. IgG anti-PA Ab response of A/J mice immunized with AVA ± 20 μg of CpG ODN. Results represent the geometric mean ± SD of serum IgG anti-PA titers 14 days after immunization ($n = 6$ independently studied mice/group in two separate experiments). * $P < 0.05$, determined by one-way ANOVA.

their ability to predict survival following STI challenge. As seen in FIGURE 2 A, TNA correlated significantly with IgG anti-PA titer ($R^2 = 0.46$, $P < 0.0001$). Two-parameter logistic regression modeling showed that IgG anti-PA titers were the superior surrogate marker for survival (FIG. 2, IgG: $R^2 = 0.64$, $P < 0.0001$; versus TNA: $R^2 = 0.36$, $P < 0.0001$). In this context, receiver operating characteristic analysis showed that IgG anti-PA titer was 97% accurate at predicting survival following anthrax challenge, whereas TNA was 91% accurate.

CpG ODN Improve the Survival of AVA-Vaccinated Guinea Pigs

To confirm and extend these findings, a guinea pig challenge model was employed. Normal guinea pigs succumb rapidly to challenge by 50 LD_{50} of encapsulated Ames strain anthrax spores (FIG. 3). Immunization and boosting with AVA alone improved their survival rate, although most animals still died from infection. By comparison, 75% of animals immunized and boosted with AVA + CpG ODN survived challenge ($P = 0.05$).

CpG ODN Increase the Speed and Titer of the IgG Anti-PA Ab Response Induced by AVA in Rhesus Macaques

As noted in the introduction, different classes of CpG ODN trigger distinct types of immune response in primates.[29,30] "D" ODN stimulate pDCs to undergo functional maturation and produce high levels of IFN-α whereas "K" ODN stimulate B cells and trigger pDCs to produce TNF-α but not IFN-α.[29,34] To determine which class of ODN might best function as an adjuvant for AVA, rhesus macaques were primed and boosted with the normal human dose of AVA plus optimized mixtures of either "K" or "D" CpG ODN. Initially, mixtures were formulated that included three different CpG motifs for each class of ODN. These mixtures were previously shown to activate PBMC from nearly all human donors tested ($n > 100$[35]).

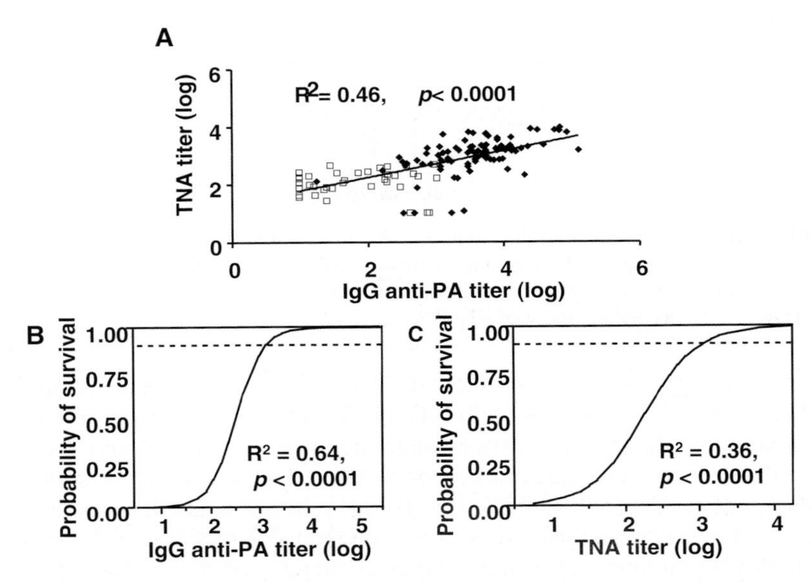

FIGURE 2. Correlation between serum Ab response and survival. Mice were immunized with AVA ± CpG ODN 1555. Two weeks post immunization, serum IgG anti-PA and TNA titers were determined, and the mice were challenged i.p. with 9×10^3 LD_{50} of STI spores. Results from four independent experiments involving a total of 130 mice are shown. (**A**) Linear regression of IgG anti-PA versus TNA titers in mice that succumbed to (□) or survived (♦) infection. (**B**) Logistic regression of survival versus IgG anti-PA titer. (**C**) Logistic regression of survival versus TNA titer.

As seen in TABLE 1, the addition of "K" ODNs to AVA induced a significantly higher IgG anti-PA response at all time points when compared to AVA alone ($P < 0.02$). Although "D" ODN also increased the IgG anti-PA response, this effect did not achieve statistical significance until animals were challenged. Thus, subsequent studies focused on the adjuvant activity of "K" class ODN.

TABLE 1. Immunogenicity of AVA combined with "K" or "D" ODN in rhesus macaques

	IgG anti-PA titer		
Immunogen	Primary	Secondary	Post challenge
AVA	9,800 ± 7,100	315,000 ± 74,000	33,300 ± 7,900
AVA + "K" ODN	25,900 ± 8,100*	624,000 ± 181,000*	95,600 ± 25,400*
AVA + "D" ODN	16,800 ± 7,600	467,000 ± 96,000	86,900 ± 23,700*

Rhesus macaques (5/group) were immunized and boosted 6 weeks later with 500 μL of AVA plus 500 μg of "K" or "D" class CpG ODN. The animals were "challenged" with 10^5 STI anthrax spores at week 26. IgG anti-PA titers were measured in the serum of each animal 3 weeks post immunization and 2 weeks post boost/challenge. Results represent the mean + SEM of these titers.

$^*P < 0.02$ vs. AVA alone.

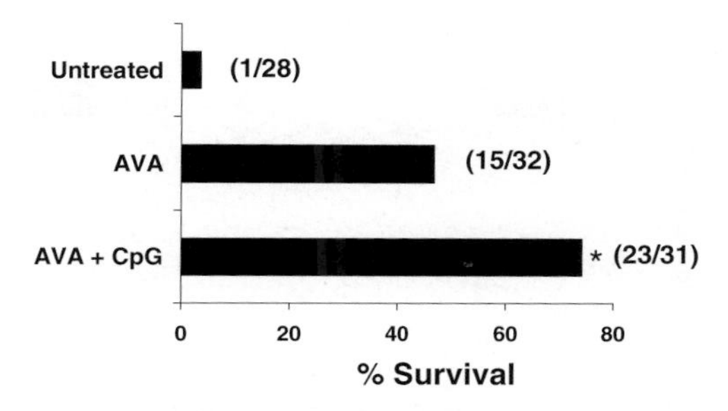

FIGURE 3. Guinea pigs were immunized on day 0 and boosted on week 4 with AVA ± CpG ODN. Six weeks later they were challenged with 50 LD_{50} *B. anthracis* Ames spores. Percent survival and the number of animals surviving/total are shown. *Significantly improved survival compared to animals immunized with AVA alone, $P = 0.05$.

The quality of a vaccine-adjuvant combination is reflected by both the avidity and titer of the Abs induced. To assess the avidity of the IgG anti-PA response, the ability of these Abs to remain bound to PA in the presence of 6 M urea was evaluated.[36,37] As expected, affinity-matured Abs induced by secondary immunization were significantly more avid than those elicited by primary immunization (FIG. 4, $P < 0.01$). The avidity of the secondary IgG anti-PA response of macaques immunized with AVA + "K" ODN was significantly higher than that of animals immunized with AVA alone (FIG. 4, $P < 0.01$).

Based on the evidence that "K" class CpG ODN improved the immunogenicity of AVA, a second experiment was conducted to compare the activity of clinical grade "K" ODN 7909 (kindly provided by Coley Pharmaceuticals) to the original "K" ODN mixture and AVA alone. As seen in FIGURE 5,

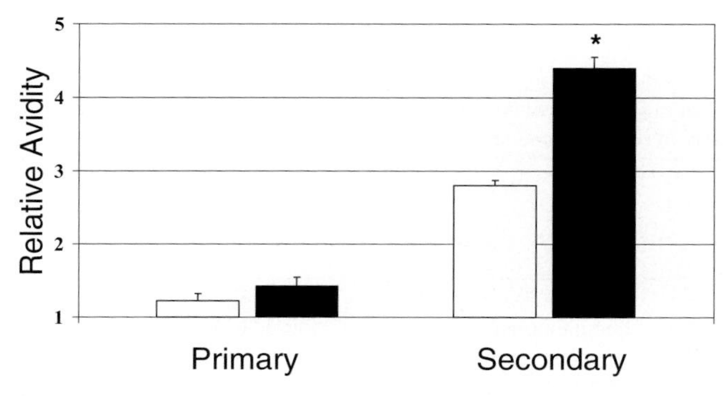

FIGURE 4. The relative avidity + SEM of the IgG anti-PA Abs in the serum of animals immunized and boosted as described in FIGURE 1 was determined by elution with 6 M urea. *Significantly greater avidity compared to animals immunized with AVA alone, $P < 0.03$.

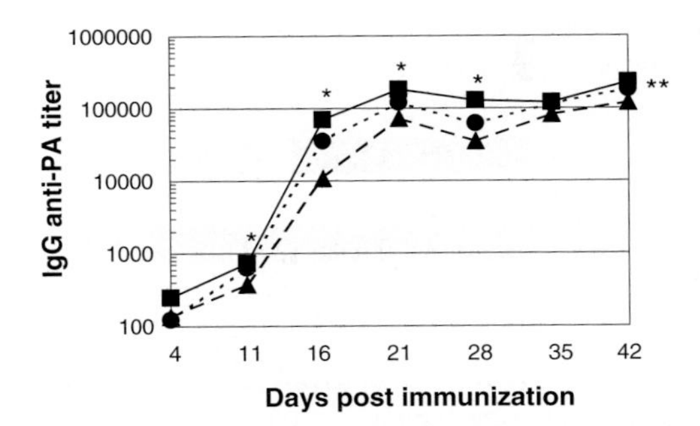

FIGURE 5. Rhesus macaques (5/group) were immunized s.c. with 0.5 mL of AVA alone (Δ) or combined with 500 μg of the "K" ODN mixture used in TABLE 1 (●) or ODN 7909 (■). IgG anti-PA titers were measured by ELISA in the serum of each animal at multiple time points post immunization. Results represent the mean + SEM of these titers. *Significantly higher IgG anti-PA titer at that time point, compared to animals immunized with AVA alone, $P < 0.05$.**Significantly higher cumulative IgG anti-PA titer compared to animals immunized with AVA alone, $P < 0.01$.

macaques immunized with AVA plus the original "K" ODN mixture mounted a stronger immune response than those vaccinated with AVA alone. ODN 7909 triggered an even higher IgG anti-PA response, exceeding that induced by AVA by greater than threefold ($P < 0.01$).

The effect of the route of administration on the immunogenicity of AVA + ODN 7909 was then examined. As seen in TABLE 2, delivering this antigen-adjuvant combination subcutaneously (the conventional route for AVA) induced a significantly stronger serum IgG anti-PA response that administering the combination intramuscularly ($P < 0.01$).

TABLE 2. Immunogenicity of AVA combined with "K" ODN administered subcutaneously or intramuscularly to rhesus macaques

	IgG anti-PA titer Days post immunization	
	D11	D16
AVA	43 ± 33	790 ± 25
AVA + "K" ODN subcutaneously	2,500 ± 1,100*	7,610 ± 2,200*
AVA + "K" ODN intramuscularly	510 ± 170	2,900 ± 850

Rhesus macaques (5-6/group) were immunized and boosted with 500 μl of AVA plus 500 μg of "K" class CpG ODN 7909. Results reflect the mean ± SD of the serum IgG anti-PA titer of serum of each independently studied animal.

*$P < 0.05$ vs. "K" intramuscularly.

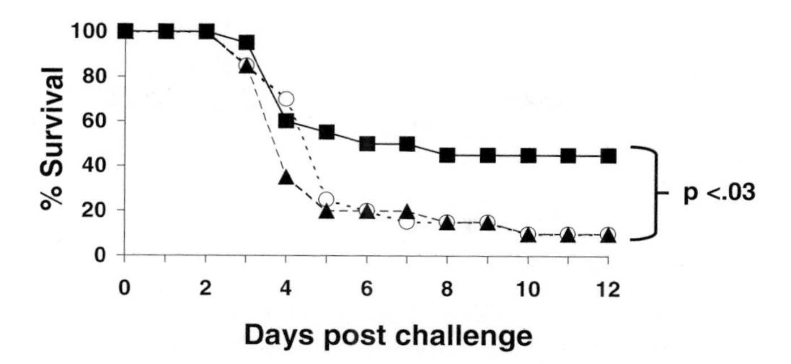

FIGURE 6. Preimmune serum (○) or serum from rhesus macaques vaccinated 11 days earlier with AVA alone (Δ) or AVA + CpG ODN 7909 (■) was pooled and injected i.p. into A/J mice (0.1 mL/recipient). The following day, mice were challenged with 30 LD_{50} of STI anthrax. Results of two independent experiments involving a combined total of 20 recipients/treatment are shown.

Protective Efficacy of the IgG Anti-PA Response in Rhesus Macaques

The critical measure of an antigen-adjuvant combination is its ability to induce protective immunity. Due to restrictions on the use of macaques in lethal anthrax challenge experiments, the ability of serum Abs from these animals to protect A/J mice (20/group) from anthrax challenge was explored. As seen in FIGURE 6, animals injected with preimmune serum or serum from AVA-immunized macaques rapidly succumbed to challenge by 30 LD_{50} of STI *B. anthracis* spores. In contrast, serum from macaques immunized with AVA plus CpG ODN protected nearly half of the recipient mice from lethal challenge ($P < 0.03$).

These differences in susceptibility correlated with the titer of toxin-neutralizing Abs present in donor serum transferred to recipient mice. Macaques immunized with AVA + ODN 7909 had on average a 17-fold higher toxin-neutralizing titer than those immunized with AVA alone (TABLE 3, $P < 0.03$). A similar difference in neutralizing Ab levels was detected in the serum of recipient mice immediately prior to the challenge (TABLE 3, $P < 0.05$). Consistent with earlier results, serum IgG anti-PA titers correlated closely with TNA ($R^2 = 0.41, P < 0.01$).

DISCUSSION

The ability of CpG DNA to stimulate immune cells expressing TLR 9 is well established. In primates, CpG ODN trigger B cells and plasmacytoid dendritic cells while in rodents additional cell types are also stimulated.[18,19] The resultant response is characterized by the functional maturation of professional antigen-presenting cells (APC), polyclonal B cell activation, and the production of Th1 and proinflammatory cytokines and chemokines.[16,17,38,39]

TABLE 3. Serum from immunized macaques transfers protection to naïve recipient mice

	Serum-neutralizing titer		
Treatment	Donor	Recipient	% Protection
Preimmunization	0 ± 0	10 ± 5	10
AVA alone	25 ± 5	43 ± 33	10
AVA + CpG ODN	434 ± 109*	2,510 ± 1,150*	45*

Rhesus macaques were immunized s.c. with 0.5 mL of AVA alone or combined with 500 μg of ODN 7909. Preimmune serum, or serum 11 days post vaccination, was pooled and injected i.p. into A/J mice (0.1 mL/recipient). The following day, mice were bled and challenged with 30 LD_{50} of STI anthrax. The serum-neutralizing titer of both donors (5/group) and recipients ($n = 20$) and percent of recipients surviving challenge are shown.

*Significantly higher than AVA group, $P < 0.05$.

Studies in mice demonstrate that CpG ODN act as immune adjuvants, significantly improving the immune response elicited by a variety of antigens.[24,25,40,41] CpG ODN have been shown to boost the immune response elicited by vaccines against influenza, measles, hepatitis B surface Ag, and tetanus toxoid by 1–3 orders of magnitude, while increasing the production of Th1 cytokines and the activity of antigen-specific CTL.[25,40–44] Due to evolutionary divergence in TLR 9 expression between species, CpG motifs optimized for activity in humans are less effective in mice.[27–29] Thus preclinical studies of CpG ODN planned for human use are best performed in nonhuman primates.[15,27–29] Studies of orangutans, aotus monkeys, and rhesus macaques showed that CpG ODN mediated a significant (several-fold) increase in the immune response elicited by the coadministered hepatitis B or heat-killed leishmania vaccines.[13–15]

This work examined the ability of various classes of CpG ODN to improve the immune response elicited by AVA (the licensed human anthrax vaccine) in mice, guinea pigs, and rhesus macaques.[32,45] Results indicate that the combination of CpG ODN plus AVA triggers a faster, higher avidity, and higher-titered immune response than AVA alone. The addition of CpG ODN to AVA significantly increased protection in all species examined (FIGS. 3 and 6). Isotype analysis of the Ab response in mice indicates that adding CpG ODN to AVA stimulates a predominantly Th1-biased immune response characterized by increased levels of IgG2a anti-PA Ab (data not shown).

The induction of IgG anti-PA Abs is a highly relevant measure of vaccine immunogenicity, since these Abs confer protection against infection.[2] Both IgG anti-PA and TNA have been proposed as surrogate markers for vaccine efficacy, although their relative merit was uncertain.[46] Resolving this issue is of considerable importance, as the decision to license future anthrax vaccines will rely on surrogate markers of protection (since conventional phase III efficacy studies cannot be conducted with biothreat pathogens). While IgG anti-PA and TNA levels both correlated with survival, two-parameter logistic regression analysis showed the former to have greater sensitivity and specificity in predicting survival (FIG. 2). From the perspective of evaluating the likely efficacy of a vaccination campaign, serum IgG anti-PA Ab titer would be useful in predicting survival following defined levels of pathogen exposure.[45]

Available data also indicate that the avidity of the secondary IgG anti-PA immune response is significantly improved by the administration of CpG ODN with AVA (FIG. 4). This effect is consistent with the documented ability of CpG ODN to promote the functional maturation of professional APC.[47] In this context, ongoing studies suggest that other methods of targeting AVA to professional APC also result in higher-titered, more avid immune responses.[45]

As with all novel therapies, the possibility of adverse side effects was considered. In previous studies, CpG ODN were safely administered to rodents and primates without adverse consequences.[15,48] In the studies reviewed herein, no serious local or systemic adverse reactions were observed in any of the macaques treated with CpG ODN plus AVA.

Vaccines targeting biothreat pathogens are typically designed for prophylactic (i.e., preexposure) use. However, vaccines capable of accelerating the development of protective immunity might be of benefit to individuals exposed to biothreat pathogens (such as workers in anthrax-contaminated buildings) and/or for "ring vaccination" of potentially exposed individuals. For these latter populations, the capacity of CpG ODN to accelerate the induction of an effective anti-PA Ab response is of particular interest. Data demonstrate that coadministration of "K" class CpG ODN with AVA generates high levels of toxin-neutralizing Ab very rapidly (exceeding AVA alone by 10-fold at 17 days post immunization, FIG. 5). Passive transfer of these serum Ab protected nearly half of the naïve mice from challenge with 30 LD_{50} of anthrax spores (FIG. 6, $P < 0.03$ versus AVA alone). These findings support the further development of CpG ODN as an adjuvant for vaccines targeting biothreat pathogens.

ACKNOWLEDGMENTS

The authors thank Dr. Lev Sirota for his assistance in analyzing these data. The assertions herein are the private ones of the authors and are not to be construed as official or as reflecting the views of USAMRIID or FDA at large. Support for this work was provided in part by the Military Interdepartmental Purchase Request # MM8926 and DARPA.

REFERENCES

1. HANNA, P. 1998. Anthrax pathogenesis and host response. Curr. Top. Microbiol. Immunol. **225:** 13–35.
2. FRIEDLANDER, A.M. & P.S. BRACHMAN. 1998. Anthrax. *In* Vaccines. S.A. Plotkin & E.A. Mortimer, Eds.: 729–739. W.B. Saunders. Philadelphia, PA.

3. LANE, H.C., J.L. MONTAGNE & A.S. FAUCI. 2001. Bioterrorism: a clear and present danger. Nat. Med. **7:** 1271–1273.
4. IVINS, B.E. & S.L. WELKOS. 1988. Recent advances in the development of an improved, human anthrax vaccine. Eur. J. Epidemiol. **4:** 12–19.
5. IVINS, B.E., S.L. WELKOS, S.F. LITTLE, *et al.*1992. Immunization against anthrax with *Bacilus anthracis* protective antigen combined with adjuvants. Infect. Immun. **60:** 662–668.
6. LITTLE, S.F. & B.E. IVINS. 1999. Molecular pathogenesis of *Bacillus anthracis* infection. Microbes. Infect. **1:** 131–139.
7. WELKOS, S., S. LITTLE, A. FRIEDLANDER, *et al.* 2001. The role of antibodies to *Bacillus anthracis* and anthrax toxin components in inhibiting the early stages of infection by anthrax spores. Microbiology **147:** 1677–1685.
8. WELKOS, S.L. & A.M. FRIEDLANDER. 1988. Comparative safety and efficacy against *Bacillus anthracis* of protective antigen and live vaccines in mice. Microb. Pathog. **5:** 127–139.
9. PITTMAN, P.R., P.H. GIBBS, T.L. CANNON & A.M. FRIEDLANDER. 2001. Anthrax vaccine: short-term safety experience in humans. Vaccine **20:** 972–978.
10. PITTMAN, P.R., G. KIM-AHN, D.Y. PIFAT, *et al.* 2002. Anthrax vaccine: immunogenicity and safety of a dose-reduction, route-change comparison study in humans. Vaccine **20:** 1412–1420.
11. GEIER, D.A. & M.R. GEIER. 2002. Anthrax vaccination and joint related adverse reactions in light of biological warfare scenarios. Clin. Exp. Rheumatol. **20:** 217–220.
12. READY, T. 2004. US soldiers refuse to fall in line with anthrax vaccination scheme. Nat. Med. **10:** 112.
13. DAVIS, H.L., I.I. SUPARTO, R.R. WEERATNA, *et al.* 2000. CpG DNA overcomes hyporesponsiveness to hepatitis B vaccine in orangutans. Vaccine **18:** 1920–1924.
14. JONES, T.R., N. OBALDIA, R.A. GRAMZINSKI, *et al.*1999. Synthetic oligodeoxynucleotides containing CpG motifs enhance immunogenic vaccine in Aotus monkeys. Vaccine **17:** 3065–3071.
15. VERTHELYI, D., R.T. KENNEY, R.A. SEDER, *et al.* 2002. CpG oligodeoxynucleotides as vaccine adjuvants in primates. J. Immunol. **168:** 1659–1663.
16. KRIEG, A.M., A. YI, S. MATSON, *et al.* 1995. CpG motifs in bacterial DNA trigger direct B-cell activation. Nature **374:** 546–548.
17. KLINMAN, D.M., A. YI, S.L. BEAUCAGE, *et al.* 1996. CpG motifs expressed by bacterial DNA rapidly induce lymphocytes to secrete IL-6, IL-12 and IFNg. Proc. Natl. Acad. Sci. USA **93:** 2879–2883.
18. TAKESHITA, F., C.A. LEIFER, I. GURSEL, *et al.* 2001. Cutting edge: role of toll-like receptor 9 in CpG DNA-induced activation of human cells. J. Immunol. **167:** 3555–3558.
19. HEMMI, H., O. TAKEUCHI, T. KAWAI, *et al.* 2000. A toll-like receptor recognizes bacterial DNA. Nature **408:** 740–745.
20. BALLAS, Z.D., W.L. RASMUSSEN & A.M. KRIEG. 1996. Induction of NK activity in murine and human cells by CpG motifs in oligodeoxynucleotides and bacterial DNA. J. Immunol. **157:** 1840–1847.
21. HALPERN, M.D., R.J. KURLANDER & D.S. PISETSKY. 1996. Bacterial DNA induces murine interferon-gamma production by stimulation of IL-12 and tumor necrosis factor-alpha. Cell. Immunol. **167:** 72–78.

22. KLINMAN, D.M., K.M. BARNHART & J. CONOVER. 1999. CpG motifs as immune adjuvants. Vaccine **17:** 19–25.
23. LIPFORD, G.B., M. BAUER, C. BLANK, *et al.* 1997. CpG-containing synthetic oligonucleotides promote B and cytotoxic T cell responses to protein antigen: a new class of vaccine adjuvants. Eur. J. Immunol. **27:** 2340–2344.
24. DAVIS, H.L. 2000. Use of CpG DNA for enhancing specific immune responses. Curr. Top. Microbiol. Immunol. **247:** 171–184.
25. MOLDOVEANU, Z., L. LOVE-HOMAN, W.Q. HUANG & A.M. KRIEG. 1998. CpG DNA, a novel immune enhancer for systemic and mucosal immunization with influenza virus. Vaccine **16:** 1216–1224.
26. KRIEG, A.M. & H.L. DAVIS. 2001. Enhancing vaccines with immune stimulatory CpG DNA. Curr. Opin. Mol. Ther. **3:** 15–24.
27. BAUER, M., K. HEEG, H. WAGNER & G.B. LIPFORD. 1999. DNA activates human immune cells through a CpG sequence-dependent manner. Immunology **97:** 699–705.
28. HARTMANN, G. & A.M. KRIEG. 2000. Mechanism and function of a newly identified CpG DNA motif in human primary B cells. J. Immunol. **164:** 944–952.
29. VERTHELYI, D., K.J. ISHII, M. GURSEL, *et al.* 2001. Human peripheral blood cells differentially recognize and respond to two distinct CpG motifs. J. Immunol. **166:** 2372–2377.
30. VOLLMER, J., R. WEERATNA, P. PAYETTE, *et al.* 2004. Characterization of three CpG oligodeoxynucleotide classes with distinct immunostimulatory activities. Eur. J. Immunol. **34:** 251–262.
31. GURSEL, I., M. GURSEL, K.J. ISHII & D.M. KLINMAN. 2001. Sterically stabilized cationic liposomes improve the uptake and immunostimulatory activity of CpG oligonucleotides. J. Immunol. **167:** 3324–3328.
32. KLINMAN, D.M., H. XIE, S.F. LITTLE, *et al.* 2004. CpG oligonucleotides improve the protective immune response induced by the anthrax vaccination of rhesus macaques. Vaccine **22:** 2881–2886.
33. AGRESTI, A. & B.A. COULL. 1996. Order-restricted tests for stratified comparisons of binomial proportions. Biometrics **52:** 1103–1111.
34. KADOWAKI, N., S. ANTONENKO & Y.J. LIU. 2001. Distinct CpG DNA and polyinosinic-polycytidylic acid ds RNA, respectively stimulate CD11c- type II dendritic precursor and CD11c+ dendritic cells to produce type I interferon. J. Immunol. **166:** 2291–2295.
35. LEIFER, C., D. VERTHELYI & D.M. KLINMAN. 2003. Human response to immunostimulatory CpG oligondeoxynucleotides. J. Immunotherapy **26:** 313–318.
36. COZON, G.J., J. FERRANDIZ, H. NEBHI, *et al.*1998. Estimation of the avidity of immunoglobulin G for routine diagnosis of chronic *Toxoplasma gondii* infection in pregnant women. Eur. J. Clin. Microbiol. Infect. Dis. **17:** 32–36.
37. EGGERS, M., U. BADER & G. ENDERS. 2000. Combination of microneutralization and avidity assays: improved diagnosis of recent primary human cytomegalovirus infection in single serum sample of second trimester pregnancy. J. Med. Virol. **60:** 324–330.
38. SPARWASSER, T., E. KOCH, R.M. VABULAS, *et al.*1998. Bacterial DNA and immunostimulatory CpG oligonucleotides trigger maturation and activation of murine dendritic cells. Eur. J. Immunol. **28:** 2045–2054.
39. ROMAN, M., E. MARTIN-OROZCO, J.S. GOODMAN, *et al.*1997. Immunostimulatory DNA sequences function as T helper-1 promoting adjuvants. Nat. Med. **3:** 849–854.

40. BRAZOLOT MILLAN, C.L., R. WEERATNA, A.M. KRIEG, *et al.* 1998. CpG DNA can induce strong Th1 humoral and cell-mediated immune responses against hepatitis B surface antigen in young mice. Proc. Natl. Acad. Sci. USA **95:** 15553–15558.
41. MCCLUSKIE, M.J. & H.L. DAVIS. 1998. CpG DNA is a potent enhancer of systemic and mucosal immune responses against hepatitis B surface antigen with intranasal administration to mice. J. Immunol. **161:** 4463–4466.
42. BRANDA, R.F., A.L. MOORE, A.R. LAFAYETTE, *et al.* 1996. Amplification of antibody production by phosphorothioate oligodeoxynucleotides. J. Lab. Clin. Med. **128:** 329–338.
43. KOVARIK, J., P. BOZZOTTI, L. LOVE-HOMAN, *et al.* 1999. CpG oligodeoxynucleotides can circumvent the Th2 polarization of neonatal responses to vaccines but may fail to fully redirect Th2 responses established by neonatal priming. J. Immunol. **162:** 1611–1617.
44. DAVIS, H.L., R. WEERANTA, T.J. WALDSCHMIDT, *et al.*1998. CpG DNA is a potent enhancer of specific immunity in mice immunized with recombinant hepatitis B surface antigen. J. Immunol. **160:** 870–876.
45. XIE, H., I. GURSEL, B.E. IVINS, *et al.* 2005. CpG oligodeoxynucleotides adsorbed onto polylactide-co-glycolide microparticles improve the immunogenicity and protective activity of the licensed anthrax vaccine. Infect. Immun. **73:** 828–833.
46. LITTLE, S.F., B.E. IVINS, P.F. FELLOWS, *et al.*2004. Defining a serological correlate of protection in rabbits for a recombinant anthrax vaccine. Vaccine **22:** 422–430.
47. GURSEL, M., D. VERTHELYI, I. GURSEL, *et al.* 2002. Differential and competitive activation of human immune cells by distinct classes of CpG oligodeoxynucleotides. J. Leukocyte Bio. **71:** 813–820.
48. KLINMAN, D.M., J. CONOVER & C. COBAN. 1999. Repeated administration of synthetic oligodeoxynucleotides expressing CpG motifs provides long-term protection against bacterial infection. Infect. Immun. **67:** 5658–5663.

Anti-VEGF Aptamer (Pegaptanib) Therapy for Ocular Vascular Diseases

EUGENE W.M. NG AND ANTHONY P. ADAMIS

(OSI) Eyetech, Inc., New York, New York 10036, USA

ABSTRACT: Vascular endothelial growth factor (VEGF) is a central regulator of both physiological and pathological angiogenesis. Pegaptanib, a 28-nucleotide RNA aptamer specific for the $VEGF_{165}$ isoform, binds to it in the extracellular space, leaving other isoforms unaffected, and inhibits such key VEGF actions as promotion of endothelial cell proliferation and survival, and vascular permeability. Pegaptanib already has been examined as a treatment for two diseases associated with ocular neovascularization, age-related macular degeneration (AMD) and diabetic macular edema (DME). Preclinical studies have shown that $VEGF_{165}$ alone mediates pathological ocular neovascularization and that its inactivation by pegaptanib inhibits the choroidal neovascularization observed in patients with neovascular AMD. In contrast, physiological vascularization, which is supported by the $VEGF_{121}$ isoform, is unaffected by this inactivation of $VEGF_{165}$. In addition, animal model studies have shown that intravitreous injection of pegaptanib can inhibit the breakdown of the blood–retinal barrier characteristic of diabetes and even can reverse this damage to some degree. These preclinical findings formed the basis for randomized controlled trials examining the efficacy of pegaptanib as a therapy for AMD and DME. The VEGF Inhibition Study in Ocular Neovascularization (VISION) trial comprising two replicate, pivotal phase 3 studies, demonstrated that intravitreous injection of pegaptanib resulted in significant clinical benefit, compared with sham injection, for all prespecified clinical end points, irrespective of patient demographics or angiographic subtype, and led to pegaptanib's approval as a treatment for AMD. A phase 2 trial has provided support for the efficacy of intravitreous pegaptanib in the treatment of DME.

KEYWORDS: pegaptanib; vascular endothelial growth factor; age-related macular degeneration; aptamer

INTRODUCTION

Aptamers are one of several oligonucleotide agents, including ribozymes, antisense RNAs, and small interfering RNAs, that have been investigated as

Address for correspondence: Anthony P. Adamis, M.D., (OSI) Eyetech, Inc. 3 Times Square, 12th Floor, New York, NY 10036. Voice: 212-824-3213; fax: 212-824-3241.
e-mail: tony.adamis@eyetech.com

Ann. N.Y. Acad. Sci. 1082: 151–171 (2006).
doi: 10.1196/annals.1348.062

therapeutic agents.[1] Analogous to monoclonal antibodies, aptamers have the ability to target extracellular molecules while, in general, other oligonucleotide agents are designed to function intracellularly. Antisense and ribozyme technologies have been under development for approximately a quarter of a century, but drug delivery issues have continued to impede the realization of their early promise (for a review see Ref. 2). In contrast, pegaptanib, the first aptamer to receive clinical approval for any disease, became available after little more than a decade of development.

Pegaptanib is a nuclease-resistant aptamer that binds with high affinity to the isoform of vascular endothelial growth factor (VEGF), which is a key regulator of both physiological and pathological angiogenesis. Currently, pegaptanib is approved for the treatment of age-related macular degeneration (AMD) and also is undergoing clinical trials as a treatment for other ocular vascular diseases, including diabetic macular edema (DME). This article focuses on the major preclinical and clinical studies confirming VEGF's pathological role in ocular diseases, together with the findings that support pegaptanib's utility as a clinical treatment for these conditions.

DEVELOPMENT OF PEGAPTANIB

The development of pegaptanib, and its ultimate application as a treatment, represents the coalescence of two independent lines of research, the development of aptamer technology on the one hand and the elucidation of the importance of VEGF as a mediator of physiological and pathological ocular angiogenesis on the other. In the late 1980s and early 1990s, VEGF already was established as a key inducer of angiogenesis (see below). Evidence for the importance of VEGF in the pathological angiogenesis emerged from preclinical studies showing that malignant tumor vascularization and growth could be inhibited by blocking VEGF, either with anti-VEGF monoclonal antibodies[3] or by expression of a viral construct bearing a dominant-negative mutant of a VEGF receptor.[4] These studies also demonstrated that the mitogenic activity released by cultured tumor cells was exerted primarily by $VEGF_{165}$,[3] the most predominant of the VEGF isoforms (see below). There was accordingly a clear impetus for the development of reagents that could serve as specific inhibitors of VEGF's pathogenic role. The concurrent development of the Selective Evaluation of Ligands by Exponential Enrichment, or SELEX, technology[5,6] for the production of aptamers provided a new approach to developing such highly specific reagents.

$VEGF_{165}$ was, accordingly, selected as a target for aptamer development by investigators at NeXstar Pharmaceuticals (formerly of Boulder, CO).[7–9] Several years of development applied toward optimizing affinity, nuclease resistance, and biological stability resulted in a 28-nucleotide, high-affinity, anti-VEGF aptamer. Endonuclease resistance was enhanced through the incorporation of 2′-fluoropyrimidines, 2′-O-methyl substitution of the majority of the purines and the addition of a 3′-3′-linked deoxythymidine terminal cap.[9] A 40 kDa polyethylene glycol moiety was added to the 5′ terminus to prolong the duration of tissue exposure.[9,10] This resulting aptamer, later named pegaptanib, has a molecular weight of approximately 49 kDa and a binding affinity for $VEGF_{165}$ of approximately 200 pM.

Preclinical experiments demonstrated that pegaptanib inhibited $VEGF_{165}$-mediated responses *in vitro* and *in vivo*. Bell *et al.*[11] showed that ^{125}I-labeled VEGF binding to cultured human umbilical vein endothelial cells (HUVECs), which express both of the cellular receptors for VEGF, VEGFR1, and VEGFR2, was inhibited by pegaptanib with an IC_{50} (concentration giving 50% inhibition) of 0.75–1.4 nM; total inhibition of VEGF binding was seen at 10 nM. Very similar results were seen with human dermal microvascular endothelial cells, showing that pegaptanib's inhibitory action was not specific to one type of endothelial cell.

Further experiments with HUVECs demonstrated that this inhibition of VEGF binding to its receptors prevented downstream signaling events, such as calcium mobilization, and the phosphorylation of both VEGFR2 and phospholipase Cγ. Most importantly, these actions were reflected in the inhibition of $VEGF_{165}$-induced proliferation of HUVECs ($IC_{50} = 1.1$ nM); in contrast, induction of HUVEC proliferation by the $VEGF_{121}$ isoform was not inhibited by pegaptanib, which is consistent with the aptamer's specificity for $VEGF_{165}$. Incubation of pegaptanib for as long as 20 h in serum-containing medium did not affect its ability to inhibit $VEGF_{165}$-mediated proliferation of HUVECs, proving the effectiveness of the measures taken to protect against nuclease degradation. Pegaptanib also inhibited VEGF activity *in vivo*, as demonstrated by the results of the Miles vascular permeability assay, which measures leakage from dermal microvessels in guinea pigs. Injection of VEGF led to a dramatic increase in vascular permeability in this assay; preincubation of VEGF with 0.1 M pegaptanib inhibited this VEGF-induced permeability by 83%.[9]

Taken together, these preclinical data demonstrated that pegaptanib inhibited two of the most important known functions of VEGF, as endothelial mitogen and as enhancer of vascular permeability. Moreover, as discussed in the following sections, an emerging body of preclinical research into ocular vascular diseases demonstrated that inhibition of these functions of VEGF could be the key to ameliorating or reversing the associated sequelae, such as visual impairment. The effects of pegaptanib were accordingly examined as part of these studies, and their elucidation not only proved essential for defining the role of VEGF in ocular vascularizing syndromes but also ultimately paved the way for pegaptanib's use as a treatment for AMD and DME.

VEGF PROMOTES PHYSIOLOGICAL AND PATHOLOGICAL NEOVASCULARIZATION

A large body of work has established VEGF as the most important of the many factors that regulate both physiological and pathological angiogenesis. VEGF was independently isolated by two groups in the 1980s, first as a tumor-secreted factor that increased vascular permeability,[12] and second as a mitogen for endothelial cells.[13] Molecular cloning revealed this factor to be a member of the VEGF-platelet-derived growth factor family. Alternative splicing generates at least four principal isoforms consisting of 121, 165, 189, and 206 amino acids. $VEGF_{165}$, the predominant isoform, is a 45 kDa glycoprotein containing a heparin-binding domain. While significant amounts are bound to the cell surface and to the extracellular matrix, much of $VEGF_{165}$ also is secreted. Of the other isoforms, $VEGF_{189}$ and $VEGF_{206}$ are highly basic and heparin binding and exist primarily as matrix-bound forms, while $VEGF_{121}$, lacking the heparin-binding domain, is freely secreted.[14]

VEGF exerts a wide range of actions of relevance to angiogenesis. In addition to being a potent endothelial mitogen, it acts as a chemoattractant to mobilize endothelial cells from the bone marrow[15–17] and as a retinal endothelial cell survival factor.[18] Moreover, it can induce the synthesis of enzymes, such as matrix metalloproteinases and a plasminogen activator that are involved in degrading the extracellular matrix, facilitating tissue penetration by proliferating vessels.[19–21]

In turn, VEGF can be released from the extracellular matrix by the action of plasmin,[14] suggesting that upregulation of plasminogen activator can result in further amplification of the effects of VEGF. A similar amplification cascade may occur through VEGF upregulation of endothelial nitric oxide synthase; this in turn leads to the release of nitric oxide, which is itself a key mediator of angiogenesis that can induce synthesis of VEGF.[22–24] VEGF expression also is upregulated by hypoxia, which may play a role in the etiology of AMD (for a review see Ref. 25). VEGF secretion by the retinal pigment epithelium, a layer of pigmented cells underlying the retina that nourishes the retinal photoreceptor cells, is an important response to ocular ischemia.[26,27] Furthermore, a variety of growth factors can modulate VEGF expression, leading to a complex interplay of interactions that regulate local VEGF levels (for a review see Ref. 28).

In addition to its proangiogenic effects, VEGF contributes to ocular vascular disease by promoting edema and inflammation. As the most potent inducer of vascular permeability known, some 50,000 times more potent than histamine,[29] VEGF is believed to be especially important in causing the exudation that underlines the loss of vision in conditions, such as AMD and DME. Several mechanisms are involved in this VEGF-mediated increase in permeability, including induction of fenestrations in the vascular endothelium,[30] dissolution of tight junctions,[31] and attraction of leukocytes that attack the endothelium.[32,33]

The key role of VEGF in pathological angiogenesis makes it an attractive target for therapeutic intervention. However, VEGF also has a wide range of beneficial effects as well. In the eye, secretion of VEGF by the retinal pigment epithelium helps to maintain the choriocapillaris (the vascular network that underlies the retina)[27] and exerts a neuroprotective action in the ischemic retina.[34] Indeed, this neuroprotective function extends to a wide variety of neurons; recent work demonstrating that VEGF modifies motor-neuron degeneration has provided a theoretical basis for a potential therapeutic application in treating such diseases as amyotrophic lateral sclerosis.[35,36] VEGF also is involved in processes as disparate as bone growth,[37,38] wound healing,[39,40] female reproductive cycling,[37,41] vasorelaxation,[42] skeletal muscle regeneration,[43] glomerulogenesis,[44] and induction of paracrine mechanisms that mediate resistance to liver damage.[45] Given the many processes in which VEGF plays a role, it becomes imperative that therapeutic strategies premised on its inactivation be administered in such a way as to minimize the possibility of systemic side effects. Complications have already been observed with the systemic administration of bevacizumab, a monoclonal antibody against VEGF that has been employed as an adjunct in chemotherapy regimens and that is associated with an enhanced risk of thromboembolic events.[46]

VEGF IN OCULAR NEOVASCULARIZATION

The cellular and molecular bases underlying mediating ocular neovascularization have been under intensive study for more than a decade, with a principal focus on the development of choroidal neovascularization (CNV), the growth of new blood vessels from the choroid. VEGF has emerged as a critical mediator of this and other manifestations of aberrant vascularization, with evidence from clinical studies and supported by preclinical models of ocular neovascularization.

VEGF is one of many growth factors that has been found in association with ocular neovascularization (for a review see Ref. 47) and it is produced by a variety of cell types in the retina, including the retinal pigment epithelium[27] as well as all major classes of retinal neurons.[48] The underlying causes for its elevation during ocular neovascularization are not known. It has been suggested, though, that the retina may be in a state of near hypoxia at all times owing to the metabolic demands of photoreceptors.[49,50] The fovea is particularly susceptible to hypoxia because of the lack of retinal vascularization in the foveal region.[49] Hypoxia elevates VEGF levels in a variety of retinal cell types,[51] and probably plays a contributory role in some syndromes.[25,49,50,52]

Clinical studies have demonstrated elevated levels of VEGF in the eyes of patients with a number of ocular diseases that are characterized by ocular neovascularization, including proliferative diabetic retinopathy,[52–54] neovascularization of the iris,[52] ischemic retinal vein occlusion,[52] retinopathy of prematurity,[55] neovascular glaucoma,[56] and DME.[57] Similarly, increased expression of VEGF was observed in maculae taken from patients with AMD when compared with control maculae.[58] Elevated expression of VEGF also was observed in the retinal pigment epithelium of choroidal neovascular membranes[59–61] and in the vascular endothelial cells and retinal pigment epithelium of CNV membranes undergoing active neovascularization. In contrast, VEGF was weakly expressed in clinically quiescent CNVs, suggesting that VEGF is important in early vascular neogenesis.[62]

Information derived from animal models provides additional support that VEGF is central to the pathogenesis of ocular vascular disease. Because there is not an animal condition that precisely mimics the natural course of AMD, experimental models have been developed that provoke neovascularization in response to induced ocular injury or by overexpressing VEGF in the eye. In a primate model, iris neovascularization developing after laser-induced occlusion of retinal veins developed in direct proportion to elevations of VEGF.[63] In other work, intravitreous injection of VEGF into monkey eyes led to iris,[64] retinal,[65] and preretinal[66] vascularization. Blood vessels in the injected eyes exhibited tortuosity, edema, capillary closure, and endothelial cell hyperplasia, reflecting the effects of VEGF in promoting endothelial cell proliferation and vascular permeability.[65,66] Overexpression of VEGF in retinal pigment epithelium in rodent eyes through the use of adenovirus vectors led to the development of new vessels that penetrated Bruch's membrane, vascular hyperpermeability, and subsequent death of overlying photoreceptors—features also characteristic of AMD.[67,68] In another approach, transgenic mice in which VEGF was selectively expressed in the retinal pigment epithelium[69] or in photoreceptors[70] through the use of tissue-specific promotors also developed abnormal ocular neovascularization.

Complementary experiments demonstrated that the development of CNV can be prevented by inactivation of VEGF. Adamis *et al.*[71] showed that intravitreal injection of a monoclonal antibody against VEGF prevented iris neovascularization induced by laser occlusion of the retinal vein in monkeys. Similar results were obtained with anti-VEGF monoclonal antibodies in a rat model of corneal neovascularization[72] and in anti-VEGF monoclonal antibody Fab fragments in a monkey model of CNV.[73] Other approaches that have been used to inactivate VEGF, with concomitant prevention of neovascularization, include intravitreal injection of pegaptanib,[74] soluble VEGF receptors,[51,75] and VEGF receptor-expressing adenoviruses.[76] Recently, VEGF expression and concomitant iridal neovascularization were inhibited by injection of an anti-VEGF antisense oligonucleotide.[77] Both retinal neovascularization and CNV in rats were reduced by administration of a novel VEGFR2 inhibitor; however, because the activity of other receptors also may be inhibited by this reagent, the significance of these findings is not clear.[78]

The Pathogenic Role of $VEGF_{165}$: Actions as an Inflammatory Cytokine

Detailed laboratory experiments as well as animal studies have shed considerable light on the cellular and molecular mechanisms by which VEGF can promote ocular neovascularization. An emerging theme of recent research is the pathogenic role of one specific VEGF isoform, $VEGF_{165}$, in promoting both pathological ocular neovascularization and the vascular injury that is characteristic of other ocular vascular diseases, such as diabetic retinopathy. Both neovascularization and vascular injury involve the influx of inflammatory cells, and the $VEGF_{165}$ isoform acts as a potent inflammatory cytokine that promotes this influx. This immune cell influx has in fact been reported in numerous clinical investigations, with macrophages being prominent in surgical specimens from patients with AMD.[49,58,79,80] These observations have provided the theoretical basis for the use of traditional anti-inflammatory agents, in particular the corticosteroid triamcinolone, as a potential treatment for diseases, such as diabetic retinopathy and AMD (for a review see Ref. 81). The most recent discoveries with VEGF, especially the role of $VEGF_{165}$, laid the groundwork for the trials of pegaptanib, an inhibitor specific for the $VEGF_{165}$ isoform, as a treatment for CNV in AMD and for the increased vascular permeability that occurs in DME.

In these preclinical studies, the most detailed description of the importance of VEGF, and particularly $VEGF_{165}$, comes from experiments examining the development of the retinal pathology associated with diabetes. VEGF promotes the inflammatory response through the upregulation of retinal synthesis of intercellular adhesion molecule-1 (ICAM-1) an adhesion molecule for leukocytes; subsequently, it was found that $VEGF_{165}$ was notably potent in mediating the inflammation. In rodent models of diabetic retinopathy, vascular damage follows leukocyte entrapment (leukostasis) in capillaries, with accompanying local nonperfusion, vascular leakage, and endothelial cell damage.[82] Several lines of evidence implicate VEGF and ICAM-1 in these processes. Both molecules are upregulated in rodent models of diabetic retinopathy,[83,84] and the relationship appears to be causal because experimental elevation of VEGF in the eyes of nondiabetic rats led to the increased expression of ICAM-1,[32,85] while the elevation of ICAM-1 in diabetic animals was inhibited by the administration of a soluble VEGF-receptor construct that binds and neutralizes VEGF.[84] Moreover, the leukostasis and vascular leakage were significantly reduced by the administration of antibodies against either ICAM-1 or its ligand, CD18[32,33,82]; similar reductions were seen in transgenic diabetic mice in which either of these molecules was genetically ablated.[86]

Further studies with rodent models established the specific pathogenic role of $VEGF_{165}$ in inducing these inflammatory processes. In a rat model of diabetic retinopathy, intravitreous injection of $VEGF_{164}$ was more proinflammatory than $VEGF_{120}$, being approximately twice as potent as $VEGF_{120}$ in mediating upregulation of ICAM-1, leukocyte adhesion, and breakdown of the blood–retinal barrier (BRB).[87] In addition, $VEGF_{164}$ was approximately threefold more effective in increasing leukocyte migration than $VEGF_{120}$[85] ($VEGF_{120/164}$ are the rodent equivalents of human $VEGF_{121/165}$). Most importantly, intravitreous injection of pegaptanib, which binds only $VEGF_{164/165}$, significantly reduced both leukostasis and BRB breakdown in this model, including reversing some of the BRB breakdown characteristics of later stages of diabetes.[87]

Further evidence supporting the pathogenic role of $VEGF_{164/165}$ has come from a mouse model of ocular neovascularization.[74] In this model, neonatal mice initially were maintained in a high-oxygen environment that was subsequently reduced to room air, creating retinal ischemia and neovascularization. This pathological neovascularization was compared to the physiological vascularization that occurs during normal retinal development, with a particular focus on the roles of the two VEGF isoforms. Both isoforms were elevated during physiological vascularization, with $VEGF_{164}$ approximately twice as abundant as $VEGF_{120}$. In the ischemia-induced pathological neovascularization, however, the $VEGF_{164}/VEGF_{120}$ ratio increased by approximately one order of magnitude, to 25.3 ± 8.7. Leukocytes were observed adhering to the leading edge of the pathological vascularization but were not seen to accompany physiological vascularization. While both the pathological and physiological vascularization were inhibited by injection of a VEGFR-Fc fusion protein that blocks the activity of both VEGF isoforms, injection of pegaptanib inhibited only the pathological vascularization (FIG. 1).[74] In another study, Ishida *et al.*[87] showed that $VEGF_{164}$-deficient transgenic mice had no impairment in normal physiological vascularization. In addition, $VEGF_{164}$ also has been found to be dispensable for VEGF-mediated neuroprotective effects in the ischemic retina, which appear to require only $VEGF_{120}$.[34]

Finally, one intriguing feature of pathological neovascularization[74] is the potential role of macrophages in enhancing this process. As described above, macrophages are commonly seen in clinical cases of pathological ocular vascularization.[58,79,80] In the retinal ischemia model system,[74] inactivation of monocyte lineage cells dramatically inhibited the formation of the pathological neovascularization, and similar findings have been reported by several other groups.[88–91] While the specific role of VEGF in attracting the monocyte/macrophage cells has not been established in these experimental models, it is noteworthy that VEGF is a chemoattractant for these cells,[92] with $VEGF_{165}$ again proving to be especially potent in this regard.[85]

Macrophage infiltration creates many possibilities for amplification of VEGF-mediated processes because they express VEGF.[58,61,72,88,91,93,94] Moreover, macrophages are an important source of other inflammatory cytokines, such as tumor necrosis factor-α (TNF-α). TNF-α can stimulate VEGF expression by the retinal pigment epithelium[95] and has been implicated as an inducer of pathological angiogenesis in the retina[96] and in other contexts.[97,98] A recent

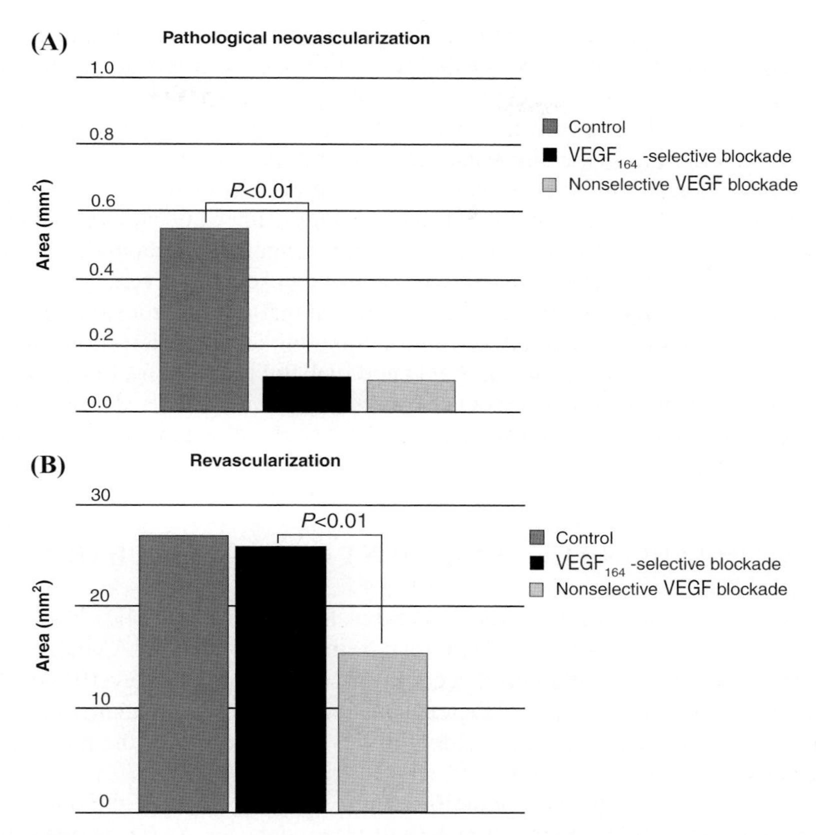

FIGURE 1. (**A**) $VEGF_{164/165}$ blockade preferentially inhibits pathological retinal neovascularization. (**B**) $VEGF_{164/165}$ blockade has no significant effect on physiological retinal vascularization. (Adapted with permission from Ref. 74.) VEGF = vascular endothelial growth factor.

case report describing partial or complete regression of CNV following systemic administration of infliximab (an anti-TNF-α monoclonal antibody) in three patients with AMD who were being treated for rheumatoid arthritis highlights a role for this cytokine in CNV.[99]

Interestingly, mice lacking monocyte chemoattractant protein-1 or its receptor, which results in impaired macrophage recruitment, were found to develop a condition remarkably similar to AMD.[100] A possible explanation for this apparent contradiction is that impaired macrophage recruitment may allow accumulation of proinflammatory molecules, such as immunoglobulins and components of the complement system that are normally removed by macrophage phagocytosis, ultimately leading to the increased production of VEGF by the retinal pigment epithelium.[100]

In conclusion, preclinical studies have provided strong evidence that ocular vascularization exhibits many attributes of an inflammatory process and that the elevation of $VEGF_{164/165}$ plays a key role in mediating this inflammation. From the clinical perspective, two groups of findings are of particular interest. First, $VEGF_{164/165}$, while especially potent in promoting the pathological vascularization in the ischemia model, is not required for physiological vascularization in the eye, such that intravitreous injection of pegaptanib inhibits only the pathological form. Second, in diabetic animals, pegaptanib is not only capable of inhibiting BRB breakdown but also can actually reverse it to some degree.[87] Taken together, these findings suggest that intravitreous injection of pegaptanib not only can be effective in treating clinical manifestations of ocular vascular conditions, such as AMD and diabetic retinopathy, but also that this treatment should not be associated with serious side effects. These expectations have indeed been borne out in clinical trials of pegaptanib in treating both AMD and DME.

PEGAPTANIB IN THE TREATMENT OF NEOVASCULAR AMD

The leading cause of blindness in the developed world is AMD,[101] and the neovascular form is estimated to be responsible for 90% of all AMD-related severe vision loss even though it accounts for only approximately 10% of the prevalence of AMD.[102] Several factors contribute to the development of AMD, but CNV, which is particularly evident in the central region of the macula, is the principal cause of vision loss.[25]

Before the approval of pegaptanib sodium injection in 2004, the only U.S. Food and Drug Administration-approved treatments for AMD involved the destruction of abnormal vessels in the eye. The options were the following: (*a*) thermal laser photocoagulation and (*b*) photodynamic therapy involving the intravenous injection of a photosensitizing compound that is then activated with a low-powered laser. Notably, thermal laser photocoagulation is highly destructive, and photodynamic therapy may only be used in a minor subset of AMD patients whose condition is characterized by a specific angiographic appearance.

The hypothesis that pegaptanib, which targets CNV, would be effective in all types of wet AMD was evaluated in two identically designed, concurrent, prospective, multicenter, randomized, double-masked, dose-ranging pivotal trials: the VEGF Inhibition Study in Ocular Neovascularization (VISION) trials. The trials were designed with the broadest possible inclusion criteria in order to reflect the patient population usually seen by clinicians. Pegaptanib sodium 0.3 mg, 1 mg, or 3 mg by intravitreous injection or sham injection was administered every 6 weeks for 48 weeks (nine injections). At week 54, patients originally randomized to receive pegaptanib were rerandomized (1:1) to continue pegaptanib for an additional 48 weeks (eight injections) or to discontinue treatment; patients originally randomized to the sham group were rerandomized (1:1:1:1:1) to either continue in the sham group, to discontinue sham, or to be treated with one of the three pegaptanib doses. Throughout the study, subconjunctival injection of anesthetic preceded all procedures, and strict ocular antisepsis was employed.[103]

The following presents the combined findings of the VISION trials. In brief, these trials demonstrated that treatment with pegaptanib reduced vision loss by approximately half in the first year and stabilized vision in the second year.

Efficacy of Pegaptanib

A total of 1208 patients were randomly assigned in the two trials; 1186 patients received at least one treatment, had baseline visual acuity assessments, and were included in year 1 efficacy analyses. At base line, demographics and ocular characteristics were similar across treatment groups. During year 1, 7545 intravitreous injections of pegaptanib and 2557 sham injections were administered. An average of 8.5 injections was administered, and approximately 90% of patients in each group completed year 1 of the study.[103]

The study's primary efficacy end point was the loss of less than 15 letters of visual acuity (approximately three lines on the study eye chart) between base line and week 54. In that time period, 55% (164/296) of patients receiving sham injections lost less than 15 letters of visual acuity compared with 70% (206/294) of patients receiving 0.3 mg of pegaptanib ($P < 0.001$), 71% (213/300) receiving 1 mg ($P < 0.001$), and 65% (193/296) receiving 3 mg ($P = 0.03$). Findings with regard to other efficacy end points were consistent with those for the primary end point; results for the 0.3-mg dose are presented in TABLE 1.[103] Doses higher than 0.3 mg were not shown to provide additional clinical benefit, and the FDA approved the 0.3-mg dose for clinical use.[103] For those receiving 0.3 mg pegaptanib and sham, average changes in visual acuity were −8 letters and −15 letters, respectively ($P < 0.0001$). There was no evidence that race, sex, iris color, baseline age, lesion size, visual acuity, or angiographic subtype precluded pegaptanib's treatment benefit.[104]

At week 54, 88% (1053/1190) of patients who received at least one treatment following base line were rerandomized, and 89% (941/1053) of those who were rerandomized were assessed at week 102. During year 2, patients rerandomized to continue therapy received an average of approximately seven injections. For year 2 analyses, data for sham patients rerandomized to either receive sham or no treatment in year 2 were combined and termed "usual care." In the combined analysis, the mean visual acuity of patients continuing 0.3 mg pegaptanib remained stable during year 2 but decreased in patients rerandomized to discontinue pegaptanib after year 1. Patients receiving usual care had the

TABLE 1. Additional year 1 efficacy end points in the VISION trials (intention-to-treat population, $n = 1186$)*[103]

End points	Pegaptanib 0.3 mg ($n = 294$)	Sham ($n = 296$)
Maintenance/gain of ≥0 letters	98 (33)	67 (23)
P value vs. sham	0.003	
Gain of ≥5 letters	64 (22)	36 (12)
P value vs. sham	0.004	
Gain of ≥10 letters	33 (11)	17 (6)
P value vs. sham	0.02	
Gain of ≥15 letters	18 (6)	6 (2)
P value vs. sham	0.04	
Loss of ≥30 letters	28 (10)	65 (22)
P value vs. sham	<0.001	
Legal blindness in study eye (visual acuity 20/200 or worse)	111 (38)	165 (56)
P value vs. sham	<0.001	

*For missing data, the "last observation carried forward" method was used.

Data are numbers of patients (%) unless otherwise noted; *P* values from the Cochran-Mantel-Haenszel test. VISION, VEGF Inhibition Study in Ocular Neovascularization.

poorest visual outcomes throughout. Patients continuing treatment with 0.3 mg pegaptanib in year 2 were less likely to lose 15 or more letters compared with those discontinuing such treatment after year 1 (Kaplan–Meier; $P < 0.05$). In descriptive analyses, percentages of patients maintaining or gaining visual acuity were higher in those receiving 2 years of 0.3 mg pegaptanib than in those receiving usual care (TABLE 2). Among patients not legally blind at base line, progression to blindness occurred in fewer patients treated with 0.3 mg pegaptanib for 2 years compared with those receiving usual care (35% versus 55%, respectively) [Data on file, (OSI) Eyetech, Inc.].

TABLE 2. Additional year 2 efficacy end points in the VISION trials (intention-to-treat population, $n = 941$)*

End points	Pegaptanib 0.3 mg –0.3 mg ($n = 133$)	Usual care ($n = 107$)
Maintenance/gain of ≥0 letters	46 (35)	28 (26)
Gain of ≥5 letters	29 (22)	15 (14)
Gain of ≥10 letters	22 (17)	6 (6)
Gain of ≥15 letters	13 (10)	4 (4)

*For missing data, the "last observation carried forward" method was used.

Data are numbers of patients (%) unless otherwise noted. VISION, VEGF (Vascular endothelial growth factor) Inhibition Study in Ocular Neovascularization.

[Data on File, (OSI) Eyetech, Inc.].

Safety of Pegaptanib

All doses of pegaptanib were found to be safe in both years 1 and 2. Most adverse events were transient, mild to moderate in intensity, and attributed to the injection procedure itself rather than to the study drug. No evidence was found of either an increased risk of potential VEGF inhibition-related adverse events or systemic toxicity.[103,105]

In year 1, less than 1% of the 7545 intravitreous injections of pegaptanib were associated with serious ocular adverse events. A total of 6 cases of retinal detachment (0.08% per injection), 5 cases of traumatic cataract (0.07% per injection), and 12 cases of endophthalmitis (0.16% per injection) were reported in year 1. Severe vision loss was seen in 1 patient with traumatic cataract and in one patient with endophthalmitis (0.01% per injection for each). Most cases of endophthalmitis followed violations of the injection preparation protocol, and 9 (75%) of the 12 patients with endophthalmitis remained in the trials for the full 54-week period.[103]

In the 374 patients treated with pegaptanib for a second year (2663 injections), there were four cases of retinal detachment (0.15% per injection) and no reports of endophthalmitis or traumatic cataract occurring during year 2. One patient with retinal detachment experienced severe vision loss (0.04% per injection).[105]

Year 1 Visual Outcomes in Patients With "Early" Lesions

Given the biology of subfoveal CNV secondary to AMD, it was hypothesized that early detection and treatment with pegaptanib may result in superior vision outcomes. To begin to test this hypothesis, an exploratory analysis of the visual outcomes at week 54 for study patients with "early" lesions was conducted. Early lesions were defined in two ways based on visual acuity, lesion characteristics, and previous treatment. Up to 20% of patients with early lesions gained $\geq$15 letters of visual acuity, and findings were consistent across analyses. New prospective studies are planned to further evaluate this theory.[106]

PEGAPTANIB IN PATIENTS WITH DME

Ocular complications, such as diabetic retinopathy and/or DME, can result from poorly controlled diabetes.[107] In diabetic retinopathy, a leading cause of blindness for individuals greater than 40 years of age,[101,108,109] vision loss occurs due to pathological processes involving increased vascular permeability and microaneurysm formation. Current treatment options are limited and include focal laser photocoagulation or removal of the vitreous humor of the eye (vitrectomy). The impact of these treatments is limited in that they do not

TABLE 3. Additional week 36 efficacy end points in the DME trial (intention-to-treat population, $n = 172$)*[114]

End points	Pegaptanib 0.3 mg ($n = 44$)	Sham ($n = 41$)
Maintenance/gain of ≥0 letters	32 (73)	21 (51)
P value vs. sham	0.02	
Gain of ≥5 letters	26 (59)	14 (34)
P value vs. sham	0.01	
Gain of ≥10 letters	15 (34)	4 (10)
P value vs. sham	0.003	
Gain of ≥15 letters	8 (18)	3 (7)
P value vs. sham	0.12	

*For missing data, the "last observation carried forward" method was used.
Data are numbers of patients (%) unless otherwise noted. DME = diabetic macular edema.

improve vision and pose a risk of additional vision loss. At best, they may help prevent further loss of vision.[110–113]

Preclinical and clinical research has confirmed that VEGF is an important mediator of DME and proliferative diabetic retinopathy.[83,87,111] Given the limitations of currently available therapy, the safety and efficacy of pegaptanib for DME were explored in a phase 2, randomized, double-masked, multicenter, dose-ranging, controlled clinical trial. Eligible patients had a best-corrected visual acuity between 20/50 and 20/320 and DME involving the center of the macula.[114] Patients received either sham injections or intravitreous pegaptanib (0.3 mg, 1 mg, or 3 mg) administered every 6 weeks for 12 weeks, with additional injections and/or focal photocoagulation as needed for another 18 weeks.

Baseline demographics and ocular characteristics were well balanced among the 172 included patients. An average of five injections was received by pegaptanib-treated patients compared with an average of 4.5 injections for those in the sham group. At week 36, median visual acuity was better in patients treated with 0.3 mg pegaptanib (20/50) compared with those receiving sham (20/63) ($P = 0.04$). Larger percentages of those receiving 0.3 mg pegaptanib maintained or gained visual acuity compared with sham (TABLE 3).[114] In addition, mean central retinal thickness decreased by 68 microns with 0.3 mg pegaptanib compared with an increase of 4 microns with sham ($P = 0.02$), and larger percentages of patients treated with 0.3 mg pegaptanib versus sham had an absolute decrease of both ≥100 microns (42% versus 16%, respectively; $P = 0.02$) and ≥75 microns (49% versus 19%, respectively; $P = 0.008$). Patients receiving pegaptanib had an approximately 50% reduction in the need for laser therapy (0.3 mg versus sham, 25% versus 48%, respectively; $P = 0.04$). Pegaptanib was well tolerated at all dose levels. Endophthalmitis occurred in 1 out of 652 injections (0.15% per injection; 0.8% per patient) and was not associated with severe vision loss. These findings are promising and suggest that pegaptanib has substantial efficacy across a broad spectrum of patients with DME.[114]

CONCLUSIONS

The benefits derived from the development and application of pegaptanib, the first aptamer approved for clinical use, foster the study of the underlying molecular basis of ocular vascular diseases as a means of rationally developing new therapies. These studies have established the central role of VEGF in ocular neovascularization. In particular, they have highlighted the role of the $VEGF_{165}$ isoform in mediating the inflammatory processes that are key to the development of both pathological neovascularization and the vascular permeability abnormalities that lead to vision loss. The clinical application of pegaptanib, an inhibitor specific for $VEGF_{165}$, emanated directly from this work.

Pivotal phase 3 trials have established the clinical efficacy of pegaptanib in treating all forms of neovascular AMD irrespective of angiographic subtype, underscoring the fact that the use of treatments based on the underlying pathophysiology of this disease may obviate the angiographic stratification of patients that has heretofore been part of clinical procedure. Moreover, phase 2 studies of pegaptanib as a treatment for DME also have been very encouraging and justify the advancement of the drug to pivotal phase 3 trials.

REFERENCES

1. BREAKER, R.R. 2004. Natural and engineered nucleic acids as tools to explore biology. Nature **432:** 838–845.
2. PERACCHI, A. 2004. Prospects for antiviral ribozymes and deoxyribozymes. Rev. Med. Virol. **14:** 47–64.
3. KIM, K.J. *et al.* 1993. Inhibition of vascular endothelial growth factor-induced angiogenesis suppresses tumour growth *in vivo*. Nature **362:** 841–844.
4. MILLAUER, B. *et al.* 1994. Glioblastoma growth inhibited *in vivo* by a dominant-negative Flk-1 mutant. Nature **367:** 576–579.
5. ELLINGTON, A.D. & J.W. SZOSTAK. 1990. *In vitro* selection of RNA molecules that bind specific ligands. Nature **346:** 818–822.
6. TUERK, C. & L. GOLD. 1990. Systematic evolution of ligands by exponential enrichment: RNA ligands to bacteriophage T4 DNA polymerase. Science **249:** 505–510.
7. JELLINEK, D. *et al.* 1994. Inhibition of receptor binding by high-affinity RNA ligands to vascular endothelial growth factor. Biochemistry **33:** 10450–10456.
8. GREEN, L.S. *et al.* 1995. Nuclease-resistant nucleic acid ligands to vascular permeability factor/vascular endothelial growth factor. Chem. Biol. **2:** 683–695.

9. RUCKMAN, J. *et al.* 1998. 2′-Fluoropyrimidine RNA-based aptamers to the 165-amino acid form of vascular endothelial growth factor (VEGF165). Inhibition of receptor binding and VEGF-induced vascular permeability through interactions requiring the exon 7-encoded domain. J. Biol. Chem. **273:** 20556–20567.
10. HEALY, J.M. *et al.* 2004. Pharmacokinetics and biodistribution of novel aptamer compositions. Pharm. Res. **21:** 2234–2246.
11. BELL, C. *et al.* 1999. Oligonucleotide NX1838 inhibits VEGF165-mediated cellular responses *in vitro*. In Vitro Cell Dev. Biol. Anim. **35:** 533–542.
12. SENGER, D.R. *et al.* 1983. Tumor cells secrete a vascular permeability factor that promotes accumulation of ascites fluid. Science **219:** 983–985.
13. LEUNG, D.W. *et al.* 1989. Vascular endothelial growth factor is a secreted angiogenic mitogen. Science **246:** 1306–1309.
14. PARK, J.E., G.A. KELLER & N. FERRARA. 1993. The vascular endothelial growth factor (VEGF) isoforms: differential deposition into the subepithelial extracellular matrix and bioactivity of extracellular matrix-bound VEGF. Mol. Biol. Cell **4:** 1317–1326.
15. ASAHARA, T. *et al.* 1999. VEGF contributes to postnatal neovascularization by mobilizing bone marrow-derived endothelial progenitor cells. Embo. J. **18:** 3964–3972.
16. LYDEN, D. *et al.* 2001. Impaired recruitment of bone-marrow-derived endothelial and hematopoietic precursor cells blocks tumor angiogenesis and growth. Nat. Med. **7:** 1194–1201.
17. CSAKY, K.G. *et al.* 2004. Recruitment of marrow-derived endothelial cells to experimental choroidal neovascularization by local expression of vascular endothelial growth factor. Exp. Eye Res. **78:** 1107–1116.
18. ALON, T. *et al.* 1995. Vascular endothelial growth factor acts as a survival factor for newly formed retinal vessels and has implications for retinopathy of prematurity. Nat. Med. **1:** 1024–1028.
19. PEPPER, M.S. *et al.* 1991. Vascular endothelial growth factor (VEGF) induces plasminogen activators and plasminogen activator inhibitor-1 in microvascular endothelial cells. Biochem. Biophys. Res. Commun. **181:** 902–906.
20. LAMOREAUX, W.J. *et al.* 1998. Vascular endothelial growth factor increases release of gelatinase A and decreases release of tissue inhibitor of metalloproteinases by microvascular endothelial cells *in vitro*. Microvasc. Res. **55:** 29–42.
21. HIRATSUKA, S. *et al.* 2002. MMP9 induction by vascular endothelial growth factor receptor-1 is involved in lung-specific metastasis. Cancer Cell **2:** 289–300.
22. PAPAPETROPOULOS, A. *et al.* 1997. Nitric oxide production contributes to the angiogenic properties of vascular endothelial growth factor in human endothelial cells. J. Clin. Invest. **100:** 3131–3139.
23. UHLMANN, S. *et al.* 2001. Direct measurement of VEGF-induced nitric oxide production by choroidal endothelial cells. Microvasc. Res. **62:** 179–189.
24. DULAK, J. *et al.* 2000. Nitric oxide induces the synthesis of vascular endothelial growth factor by rat vascular smooth muscle cells. Arterioscler. Thromb. Vasc. Biol. **20:** 659–666.
25. AMBATI, J. *et al.* 2003. Age-related macular degeneration: etiology, pathogenesis, and therapeutic strategies. Surv. Ophthalmol. **48:** 257–293.
26. AIELLO, L.P. *et al.* 1995. Hypoxic regulation of vascular endothelial growth factor in retinal cells. Arch. Ophthalmol. **113:** 1538–1544.

27. BLAAUWGEERS, H.G. *et al.* 1999. Polarized vascular endothelial growth factor secretion by human retinal pigment epithelium and localization of vascular endothelial growth factor receptors on the inner choriocapillaris. Evidence for a trophic paracrine relation. Am. J. Pathol. **155:** 421–428.
28. FERRARA, N. 2004. Vascular endothelial growth factor: basic science and clinical progress. Endocr. Rev. **25:** 581–611.
29. SENGER, D.R. *et al.* 1990. Purification and NH2-terminal amino acid sequence of guinea pig tumor-secreted vascular permeability factor. Cancer Res. **50:** 1774–1778.
30. ROBERTS, W.G. & G.E. PALADE. 1997. Neovasculature induced by vascular endothelial growth factor is fenestrated. Cancer Res. **57:** 765–772.
31. ANTONETTI, D.A. *et al.* 1999. Vascular endothelial growth factor induces rapid phosphorylation of tight junction proteins occludin and zonula occluden 1. A potential mechanism for vascular permeability in diabetic retinopathy and tumors. J. Biol. Chem. **274:** 23463–23467.
32. MIYAMOTO, K. *et al.* 2000. Vascular endothelial growth factor (VEGF)-induced retinal vascular permeability is mediated by intercellular adhesion molecule-1 (ICAM-1). Am. J. Pathol. **156:** 1733–1739.
33. JOUSSEN, A.M. *et al.* 2001. Leukocyte-mediated endothelial cell injury and death in the diabetic retina. Am. J. Pathol. **158:** 147–152.
34. SHIMA, D.T. *et al.* 2004. VEGF-mediated neuroprotection in ischemic retina.[E-abstract]. Invest. Ophthalmol. Vis. Sci. **45:** 3270.
35. STORKEBAUM, E. *et al.* 2005. Treatment of motoneuron degeneration by intracerebroventricular delivery of VEGF in a rat model of ALS. Nat. Neurosci. **8:** 85–92.
36. STORKEBAUM, E., D. LAMBRECHTS & P. CARMELIET. 2004. VEGF: once regarded as a specific angiogenic factor, now implicated in neuroprotection. Bioessays **26:** 943–954.
37. RYAN, A.M. *et al.* 1999. Preclinical safety evaluation of rhuMAbVEGF, an antiangiogenic humanized monoclonal antibody. Toxicol. Pathol. **27:** 78–86.
38. GERBER, H.P. *et al.* 1999. VEGF couples hypertrophic cartilage remodeling, ossification and angiogenesis during endochondral bone formation. Nat. Med. **5:** 623–628.
39. NISSEN, N.N. *et al.* 1998. Vascular endothelial growth factor mediates angiogenic activity during the proliferative phase of wound healing. Am. J. Pathol. **152:** 1445–1452.
40. DEODATO, B. *et al.* 2002. Recombinant AAV vector encoding human VEGF165 enhances wound healing. Gene Ther. **9:** 777–785.
41. FRASER, H.M. *et al.* 2005. Single injections of vascular endothelial growth factor trap block ovulation in the macaque and produce a prolonged, dose-related suppression of ovarian function. J. Clin. Endocrinol. Metab. **90:** 1114–1122.
42. LIU, M.H. *et al.* 2002. Vascular endothelial growth factor-mediated, endothelium-dependent relaxation in human internal mammary artery. Ann. Thorac. Surg. **73:** 819–824.
43. ARSIC, N. *et al.* 2004. Vascular endothelial growth factor stimulates skeletal muscle regeneration *in vivo*. Mol. Ther. **10:** 844–854.
44. KITAMOTO, Y., H. TOKUNAGA & K. TOMITA. 1997. Vascular endothelial growth factor is an essential molecule for mouse kidney development: glomerulogenesis and nephrogenesis. J. Clin. Invest. **99:** 2351–2357.

45. LeCouter, J. *et al.* 2003. Angiogenesis-independent endothelial protection of liver: role of VEGFR-1. Science **299:** 890–893.
46. Skillings, J.R. *et al.* 2005. Arterial thromboembolic events (ATEs) in a pooled analysis of 5 randomized, controlled trials (RCTs) of bevacizumab (BV) with chemotherapy [Meeting Abstracts]. J. Clin. Oncol. **23:** 3019.
47. Schlingemann, R.O. 2004. Role of growth factors and the wound healing response in age-related macular degeneration. Graefes Arch. Clin. Exp. Ophthalmol. **242:** 91–101.
48. Famiglietti, E.V. *et al.* 2003. Immunocytochemical localization of vascular endothelial growth factor in neurons and glial cells of human retina. Brain Res. **969:** 195–204.
49. Penfold, P.L. *et al.* 2001. Immunological and aetiological aspects of macular degeneration. Prog. Retin. Eye Res. **20:** 385–414.
50. Arden, G.B. & J.E. Wolf. 2003. Differential effects of light and alcohol on the electro-oculographic responses of patients with age-related macular disease. Invest. Ophthalmol. Vis. Sci. **44:** 3226–3232.
51. Aiello, L.P. *et al.* 1995. Suppression of retinal neovascularization *in vivo* by inhibition of vascular endothelial growth factor (VEGF) using soluble VEGF-receptor chimeric proteins. Proc. Natl. Acad. Sci. USA **92:** 10457–10461.
52. Aiello, L.P. *et al.* 1994. Vascular endothelial growth factor in ocular fluid of patients with diabetic retinopathy and other retinal disorders. N. Engl. J. Med. **331:** 1480–1487.
53. Adamis, A.P. *et al.* 1994. Increased vascular endothelial growth factor levels in the vitreous of eyes with proliferative diabetic retinopathy. Am. J. Ophthalmol. **118:** 445–450.
54. Watanabe, D. *et al.* 2005. Vitreous levels of angiopoietin 2 and vascular endothelial growth factor in patients with proliferative diabetic retinopathy. Am. J. Ophthalmol. **139:** 476–481.
55. Lashkari, K. *et al.* 2000. Vascular endothelial growth factor and hepatocyte growth factor levels are differentially elevated in patients with advanced retinopathy of prematurity. Am. J. Pathol. **156:** 1337–1344.
56. Tripathi, R.C. *et al.* 1998. Increased level of vascular endothelial growth factor in aqueous humor of patients with neovascular glaucoma. Ophthalmology **105:** 232–237.
57. Funatsu, H. *et al.* 2002. Angiotensin II and vascular endothelial growth factor in the vitreous fluid of patients with diabetic macular edema and other retinal disorders. Am. J. Ophthalmol. **133:** 537–543.
58. Kliffen, M. *et al.* 1997. Increased expression of angiogenic growth factors in age-related maculopathy. Br. J. Ophthalmol. **81:** 154–162.
59. Frank, R.N. *et al.* 1996. Basic fibroblast growth factor and vascular endothelial growth factor are present in epiretinal and choroidal neovascular membranes. Am. J. Ophthalmol. **122:** 393–403.
60. Lopez, P.F. *et al.* 1996. Transdifferentiated retinal pigment epithelial cells are immunoreactive for vascular endothelial growth factor in surgically excised age-related macular degeneration-related choroidal neovascular membranes. Invest. Ophthalmol. Vis. Sci. **37:** 855–868.
61. Grossniklaus, H.E. *et al.* 2002. Macrophage and retinal pigment epithelium expression of angiogenic cytokines in choroidal neovascularization. Mol. Vis. **8:** 119–126.

62. MATSUOKA, M. *et al.* 2004. Expression of pigment epithelium derived factor and vascular endothelial growth factor in choroidal neovascular membranes and polypoidal choroidal vasculopathy. Br. J. Ophthalmol. **88:** 809–815.
63. MILLER, J.W. *et al.* 1994. Vascular endothelial growth factor/vascular permeability factor is temporally and spatially correlated with ocular angiogenesis in a primate model. Am. J. Pathol. **145:** 574–584.
64. TOLENTINO, M.J. *et al.* 1996. Vascular endothelial growth factor is sufficient to produce iris neovascularization and neovascular glaucoma in a nonhuman primate. Arch. Ophthalmol. **114:** 964–970.
65. TOLENTINO, M.J. *et al.* 1996. Intravitreous injections of vascular endothelial growth factor produce retinal ischemia and microangiopathy in an adult primate. Ophthalmology **103:** 1820–1828.
66. TOLENTINO, M.J. *et al.* 2002. Pathologic features of vascular endothelial growth factor-induced retinopathy in the nonhuman primate. Am. J. Ophthalmol. **133:** 373–385.
67. BAFFI, J. *et al.* 2000. Choroidal neovascularization in the rat induced by adenovirus mediated expression of vascular endothelial growth factor. Invest. Ophthalmol. Vis. Sci. **41:** 3582–3589.
68. SPILSBURY, K. *et al.* 2000. Overexpression of vascular endothelial growth factor (VEGF) in the retinal pigment epithelium leads to the development of choroidal neovascularization. Am. J. Pathol. **157:** 135–144.
69. SCHWESINGER, C. *et al.* 2001. Intrachoroidal neovascularization in transgenic mice overexpressing vascular endothelial growth factor in the retinal pigment epithelium. Am. J. Pathol. **158:** 1161–1172.
70. OHNO-MATSUI, K. *et al.* 2002. Inducible expression of vascular endothelial growth factor in adult mice causes severe proliferative retinopathy and retinal detachment. Am. J. Pathol. **160:** 711–719.
71. ADAMIS, A.P. *et al.* 1996. Inhibition of vascular endothelial growth factor prevents retinal ischemia-associated iris neovascularization in a nonhuman primate. Arch. Ophthalmol. **114:** 66–71.
72. AMANO, S. *et al.* 1998. Requirement for vascular endothelial growth factor in wound- and inflammation-related corneal neovascularization. Invest. Ophthalmol. Vis. Sci. **39:** 18–22.
73. KRZYSTOLIK, M.G. *et al.* 2002. Prevention of experimental choroidal neovascularization with intravitreal anti-vascular endothelial growth factor antibody fragment. Arch. Ophthalmol. **120:** 338–346.
74. ISHIDA, S. *et al.* 2003. $VEGF_{164}$-mediated inflammation is required for pathological, but not physiological, ischemia-induced retinal neovascularization. J. Exp. Med. **198:** 483–489.
75. AGOSTINI, H. *et al.* 2005. A single local injection of recombinant VEGF receptor 2 but not of Tie2 inhibits retinal neovascularization in the mouse. Curr. Eye Res. **30:** 249–257.
76. HONDA, M. *et al.* 2000. Experimental subretinal neovascularization is inhibited by adenovirus-mediated soluble VEGF/flt-1 receptor gene transfection: a role of VEGF and possible treatment for SRN in age-related macular degeneration. Gene Ther. **7:** 978–985.
77. BHISITKUL, R.B. *et al.* 2005. An antisense oligodeoxynucleotide against vascular endothelial growth factor in a nonhuman primate model of iris neovascularization. Arch. Ophthalmol. **123:** 214–219.

78. Kinose, F. *et al.* 2005. Inhibition of retinal and choroidal neovascularization by a novel KDR kinase inhibitor. Mol. Vis. **11:** 366–373.
79. Penfold, P.L., M.C. Killingsworth & S.H. Sarks. 1985. Senile macular degeneration: the involvement of immunocompetent cells. Graefes Arch. Clin. Exp. Ophthalmol. **223:** 69–76.
80. Grossniklaus, H.E. *et al.* 2000. Correlation of histologic 2-dimensional reconstruction and confocal scanning laser microscopic imaging of choroidal neovascularization in eyes with age-related maculopathy. Arch. Ophthalmol. **118:** 625–629.
81. Ciulla, T.A. *et al.* 2004. Corticosteroids in posterior segment disease: an update on new delivery systems and new indications. Curr. Opin. Ophthalmol. **15:** 211–220.
82. Miyamoto, K. *et al.* 1999. Prevention of leukostasis and vascular leakage in streptozotocin-induced diabetic retinopathy via intercellular adhesion molecule-1 inhibition. Proc. Natl. Acad. Sci. USA **96:** 10836–10841.
83. Qaum, T. *et al.* 2001. VEGF-initiated blood-retinal barrier breakdown in early diabetes. Invest. Ophthalmol. Vis. Sci. **42:** 2408–2413.
84. Joussen, A.M. *et al.* 2002. Retinal vascular endothelial growth factor induces intercellular adhesion molecule-1 and endothelial nitric oxide synthase expression and initiates early diabetic retinal leukocyte adhesion *in vivo*. Am. J. Pathol. **160:** 501–509.
85. Usui, T. *et al.* 2004. $VEGF_{164(165)}$ as the pathological isoform: differential leukocyte and endothelial responses through VEGFR1 and VEGFR2. Invest. Ophthalmol. Vis. Sci. **45:** 368–374.
86. Joussen, A.M. *et al.* 2004. A central role for inflammation in the pathogenesis of diabetic retinopathy. FASEB J. **18:** 1450–1452.
87. Ishida, S. *et al.* 2003. $VEGF_{164}$ is proinflammatory in the diabetic retina. Invest. Ophthalmol. Vis. Sci. **44:** 2155–2162.
88. Sakurai, E. *et al.* 2003. Macrophage depletion inhibits experimental choroidal neovascularization. Invest. Ophthalmol. Vis. Sci. **44:** 3578–3585.
89. Ishida, S. *et al.* 2003. Leukocytes mediate retinal vascular remodeling during development and vaso-obliteration in disease. Nat. Med. **9:** 781–788.
90. Espinosa-Heidmann, D.G. *et al.* 2003. Macrophage depletion diminishes lesion size and severity in experimental choroidal neovascularization. Invest. Ophthalmol. Vis. Sci. **44:** 3586–3592.
91. Tsutsumi, C. *et al.* 2003. The critical role of ocular-infiltrating macrophages in the development of choroidal neovascularization. J. Leukoc. Biol. **74:** 25–32.
92. Barleon, B. *et al.* 1996. Migration of human monocytes in response to vascular endothelial growth factor (VEGF) is mediated via the VEGF receptor flt-1. Blood **87:** 3336–3343.
93. Harmey, J.H. *et al.* 1998. Regulation of macrophage production of vascular endothelial growth factor (VEGF) by hypoxia and transforming growth factor beta-1. Ann. Surg. Oncol. **5:** 271–278.
94. Naug, H.L. *et al.* 2000. Vitreal macrophages express vascular endothelial growth factor in oxygen-induced retinopathy. Clin. Exp. Ophthalmol. **28:** 48–52.
95. Oh, H. *et al.* 1999. The potential angiogenic role of macrophages in the formation of choroidal neovascular membranes. Invest. Ophthalmol. Vis. Sci. **40:** 1891–1898.

96. GARDINER, T.A. *et al.* 2005. Inhibition of tumor necrosis factor-alpha improves physiological angiogenesis and reduces pathological neovascularization in ischemic retinopathy. Am. J. Pathol. **166:** 637–644.
97. LEIBOVICH, S.J. *et al.* 1987. Macrophage-induced angiogenesis is mediated by tumour necrosis factor-alpha. Nature **329:** 630–632.
98. FRATER-SCHRODER, M. *et al.* 1987. Tumor necrosis factor type alpha, a potent inhibitor of endothelial cell growth *in vitro*, is angiogenic *in vivo*. Proc. Natl. Acad. Sci. USA **84:** 5277–5281.
99. MARKOMICHELAKIS, N.N., P.G. THEODOSSIADIS & P.P. SFIKAKIS. 2005. Regression of neovascular age-related macular degeneration following infliximab therapy. Am. J. Ophthalmol. **139:** 537–540.
100. AMBATI, J. *et al.* 2003. An animal model of age-related macular degeneration in senescent Ccl-2- or Ccr-2-deficient mice. Nat. Med. **9:** 1390–1397.
101. CONGDON, N. *et al.* 2004. Causes and prevalence of visual impairment among adults in the United States. Arch. Ophthalmol. **122:** 477–485.
102. FERRIS, F.L. 3rd, S.L. FINE & L. HYMAN. 1984. Age-related macular degeneration and blindness due to neovascular maculopathy. Arch. Ophthalmol. **102:** 1640–1642.
103. GRAGOUDAS, E.S. *et al.* 2004. Pegaptanib for neovascular age-related macular degeneration. N. Engl. J. Med. **351:** 2805–2816.
104. NG, E.W. & A.P. ADAMIS. 2005. Targeting angiogenesis, the underlying disorder in neovascular age-related macular degeneration. Can. J. Ophthalmol. **40:** 352–368.
105. MIELER, W., for theVEGF Inhibition Study in Ocular Neovascularization (VISION) Clinical Trial Group. 2005. Safety evaluation of second year treatment of age-related macular degeneration with pegaptanib sodium (Macugen^{TM}): VEGF Inhibition Study in Ocular Neovascularization (VISION). Invest. Ophthalmol. Vis. Sci. **46:** 1380.
106. GONZALES, C.R., for the Macugen Diabetic Retinopathy Study Group. 2005. Enhanced efficacy associated with early treatment of neovascular age-related macular degeneration with pegaptanib sodium: an exploratory analysis. Retina **25:** 815–827.
107. CIULLA, T.A., A.G. AMADOR & B. ZINMAN. 2003. Diabetic retinopathy and diabetic macular edema: pathophysiology, screening, and novel therapies. Diabetes Care **26:** 2653–2664.
108. KEMPEN, J.H. *et al.* 2004. The prevalence of diabetic retinopathy among adults in the United States. Arch. Ophthalmol. **122:** 552–563.
109. RESNIKOFF, S. *et al.* 2004. Global data on visual impairment in the year 2002. Bull. World Health Organ. **82:** 844–851.
110. AIELLO, L.P. *et al.* 1998. Diabetic retinopathy. Diabetes Care **21:** 143–156.
111. AIELLO, L.P. 2003. Perspectives on diabetic retinopathy. Am. J. Ophthalmol. **136:** 122–135.
112. SMIDDY, W.E. & H.W. FLYNN, JR. 1999. Vitrectomy in the management of diabetic retinopathy. Surv. Ophthalmol. **43:** 491–507.
113. YAMAMOTO, T. *et al.* 2003. Early postoperative retinal thickness changes and complications after vitrectomy for diabetic macular edema. Am. J. Ophthalmol. **135:** 14–19.
114. MACUGEN DIABETIC RETINOPATHY STUDY GROUP. 2005. A phase II randomized double-masked trial of pegaptanib, an anti-vascular endothelial growth factor aptamer, for diabetic macular edema. Ophthalmology **112:** 1748–1758.

RNAi in Combination with a Ribozyme and TAR Decoy for Treatment of HIV Infection in Hematopoietic Cell Gene Therapy

MINGJIE LI,[a,b] HAITANG LI,[a] AND JOHN J. ROSSI[a]

[a]*Division of Molecular Biology, Beckman Research Institute of the City of Hope, Duarte, California 91010, USA*

[b]*Department of Neurology, Washington University School of Medicine, St. Louis, Missouri 63110, USA*

ABSTRACT: Combinatorial therapies for the treatment of HIV infection have changed the course of the AIDS epidemic in developed nations where the antiviral drug combinations are readily available. Despite this progress, there are many problems associated with chemotherapy for AIDS including toxicities and emergence of viral mutants resistant to the drugs. Our goal has been the development of a hematopoietic gene therapy treatment for HIV infection. Like chemotherapy, gene therapy for treatment of HIV infection should be used combinatorially. We have thus combined three different inhibitory genes for treatment of HIV infection into a single lentiviral vector backbone. The inhibitory agents engage RNAi via a short hairpin RNA targeting HIV tat/rev mRNAs, a nucleolar localizing decoy that binds and sequesters the HIV Tat protein, and a ribozyme that cleaves and downregulates the CCR5 chemokine receptor used by HIV for cellular entry. This triple combination has proven to be highly effective for inhibiting HIV replication in primary hematopoietic cells, and is currently on track for human clinical application.

KEYWORDS: RNAi; shRNA; TAR decoy; ribozyme; HIV; AIDS; oligonucleotides

INTRODUCTION

During the past decade remarkable progress has been made in the treatment of HIV infection by combining the use of antivirals that interact with different viral targets. Drug resistance has been prolonged by the use of combinations

Address for correspondence: John J. Rossi, Division of Molecular Biology, Beckman Research Institute of the City of Hope, 1450 E. Duarte Road, Duarte, CA 91010. Voice: 626-301-8360; fax: 626-301-8271.
e-mail: jrossi@coh.org

Ann. N.Y. Acad. Sci. 1082: 172–179 (2006). © 2006 New York Academy of Sciences.
doi: 10.1196/annals.1348.006

of RT, protease, and entry inhibitors, or highly active antiretroviral therapy (HAART).[1] Despite the great progress, there is a continual need for the development of new drugs to combat the emergence of resistance mutations and to ameliorate the costs and toxicities of certain drug combinations. Unfortunately, nearly all HIV-infected individuals on HAART will need to maintain their medications for the entirety of their lives, resulting in considerable expense[2] and sometimes the toxic side effects of these drugs.[3,4] One recent report documents that in the age of HAART the majority of emergency room visits by HIV-infected individuals has shifted from opportunistic infections to treatments for antiretroviral drug-related toxicities.[4] There still remains the need for new drug development. The cost of developing a single chemotherapeutic drug is extremely high, often in the hundreds of millions of dollars. The consequences of this are that new and effective drugs are relatively slow in development and application. Given the emergence of viral mutants resistant to multiple drugs, the slow rate of new drug development, the cost per patient of HAART therapy, and the related toxicities, there is clearly a need for alternative approaches to treating HIV infection.

GENE THERAPY APPROACHES FOR TREATMENT OF HIV-1

The idea of using gene therapy for the treatment of HIV infection is certainly not new. Several clinical trials involving gene therapy of T lymphocytes or hematopoietic stem cells (HSCs) have been initiated over the past 12 years.[5–7] To date, there have been only limited reports of efficacy since most of the trials have been either safety studies or proof of principal. A limitation for HSC-based gene therapy has been inefficient transduction of transgenes into pluripotent hematopoietic progenitor cells, resulting in a very small population of protected cells.[5–7] All previous clinical trials in HSCs have utilized murine-based retroviral vectors to deliver to the therapeutic genes. Since these vectors are best at transducing actively dividing cells, they most often transduce committed progenitor cells, which are not self-renewing. Within the past several years there has been a tremendous amount of progress in the development of lentiviral vectors for gene delivery.[8–16] These vectors have the distinct advantage of being capable of infecting cells in G_0, or slowly dividing cell populations. Self-renewing pluripotent stem cells fall into the category of nondividing or slowly dividing cells. Although no lentiviral vector-mediated transductions of HSCs have been used clinically to date, VIRxSYS has an ongoing clinical trial using lentiviral vector transduced $CD4^+$ lymphocytes.[17] This trial represents the first FDA approved use of a lentiviral vector for gene therapy.

To date, all of the AIDS gene therapy trials have relied upon a single gene as the inhibitory agent. This, of course, will not be a reasonable approach for long-term therapy due to the same problems of viral resistance mutations that take place during HAART. Research conducted over the past several years in

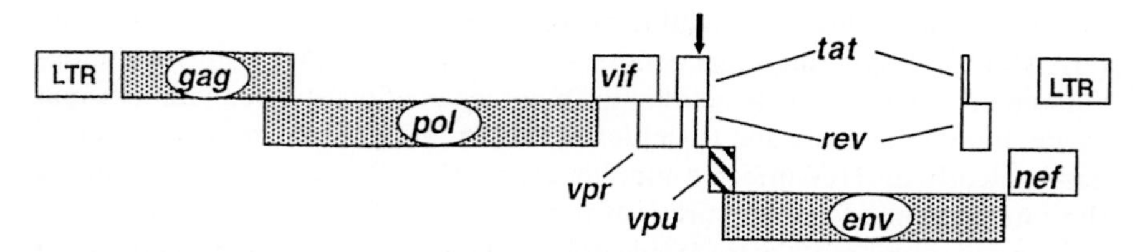

FIGURE 1. Transcript map of the HIV genome and location of shRNA target sequence in overlapping tat/rev exons.

our lab has led to the development of a novel combination of antiviral genes that target different genes in the virus as well as the co-receptor CCR5.[7,18,19] The end result of this work are two planned human clinical trials of HIV-infected individuals using a combination of three antiviral genes delivered by a lentiviral vector.[20,21] These genes include a short hairpin RNA (shRNA) targeting an HIV tat/rev common exon,[18] a nucleolar localizing TAR decoy,[22] and a ribozyme that degrades the CCR5 mRNA.[23]

RNAi AND HIV TREATMENT

Of all the antiviral agents we have tested, the most potent is RNA interference (RNAi). Our initial studies of RNAi-mediated inhibition of HIV-1 were conducted by co-transfecting plasmids that expressed separate sense and anti-sense 21-base long transcripts targeting HIV genes along with HIV-1 proviral DNA into human HEK293 cells. This resulted in over four logs of inhibition of HIV p24 antigen production relative to the controls.[24] The siRNAs used in this study targeted a common exon shared by *tat* and *rev* or the *rev* transcript alone (FIG. 1). These studies prompted us to investigate those targets in the HIV genome that were most susceptible to RNAi, and a series of shRNAs targeting various sites along the HIV genome were synthesized and cotransfected with HIV proviral DNA in HEK293 cells to find the most potent inhibitors.[25] Our results demonstrated that a shRNA targeting the common exon shared by *tat* and *rev* was the best target. This result led to further testing of this shRNA in T lymphocytes and in hematopoietic progenitor cell ($CD34^+$)-derived monocytes and macrophages.

Despite the potent inhibition mediated by the shRNA target tat/rev, the genetic variability of HIV has resulted in resistant mutants harboring one or more point mutations in the region complementary to the shRNA (FIG. 2). These results have prompted us to investigate combinatorial approaches that include a shRNA along with other RNA-based antivirals.

To date, all of the AIDS gene therapy trials have relied upon a single gene as the inhibitory agent. This, of course, will not be a reasonable approach for long-term therapy due to the same problems of viral resistance mutations that

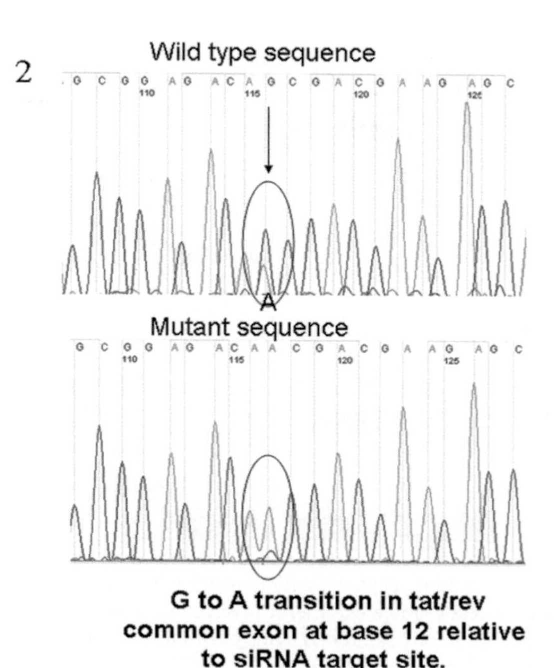

FIGURE 2. Point mutant arising in target sequence of anti-tat/rev shRNA following long-term incubation of HIV with CEM cells transduced with anti-tat/rev shRNA expressing lentiviral vector. The position of the mutation relative to the 5′ end of the shRNA is indicated. The mutant virus was amplified by RT-PCR from the supernatant of cells transduced with the above construct. The amplified DNA was cloned into a TA cloning vector and subjected to DNA sequencing. The upper panel shows a mixture of the mutant and wild-type sequences in the supernatant from the shRNA-expressing cells. The bottom panel shows the complete dominance of the mutant following rechallenge of anti-tat/rev shRNA-expressing cells.

take place during HAART. Research conducted over the past several years in our lab has led to the development of a novel combination of antiviral genes that target different genes in the virus as well as the co-receptor CCR5.[7,18,19] These genes include a shRNA targeting an HIV tat/rev common exon,[18] a nucleolar localizing TAR decoy[22] and a ribozyme that degrades the CCR5 mRNA[23,24] (FIG. 3).

Our goal has been to combine the various HIV-based inhibitors into a single lentiviral vector for the transduction of hematopoietic progenitor cells, which in turn will be reinfused into HIV patients. The three RNA-based inhibitors of FIGURE 3 have been successfully combined into a single lentiviral vector, and have been tested for anti-HIV efficacy in $CD34^+$-derived monocytes.[19] The rationale for the combination is that three different mechanisms for inhibition would circumvent viral resistance by providing multiple modes of inhibition. The CCR5 ribozyme was shown to effectively downregulate the mRNA

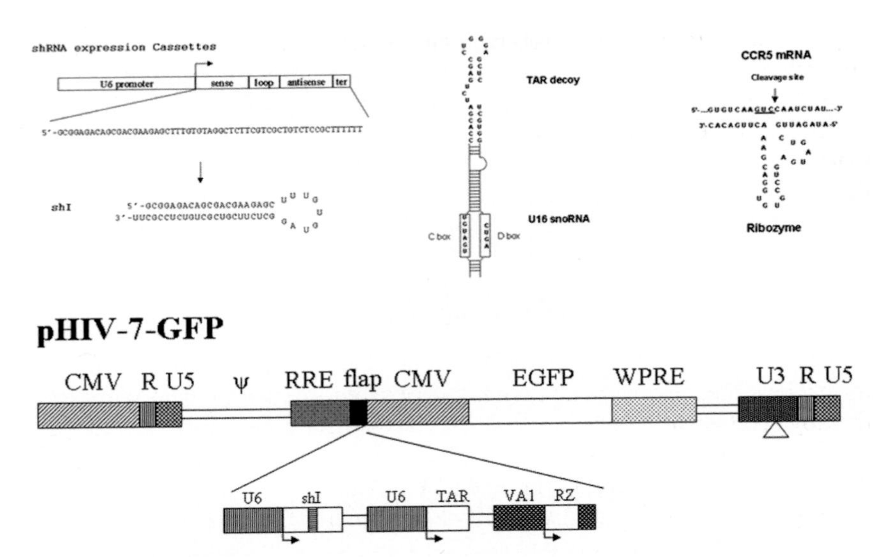

FIGURE 3. Combination of anti-HIV inhibitory RNAs. (**A**) A shRNA targeting tat/rev, a nucleolar localizing decoy, and a hammerhead ribozyme targeting the CCR5 chemokine receptor are shown. (**B**) Lentiviral vector with positions of Pol III transcription units of anti-HIV genes shown above. The shRNA and U16TAR decoy are transcribed with the U6 promoter, and the ribozyme is part of a chimeric VA1 transcript.[23]

encoding the CCR5 chemokine receptor,[18] which is used as a co-receptor for HIV entry. The TAR decoy is embedded in a nucleolar RNA and localizes to the nucleolus where it sequesters Tat from the HIV promotor.[22] The shRNA

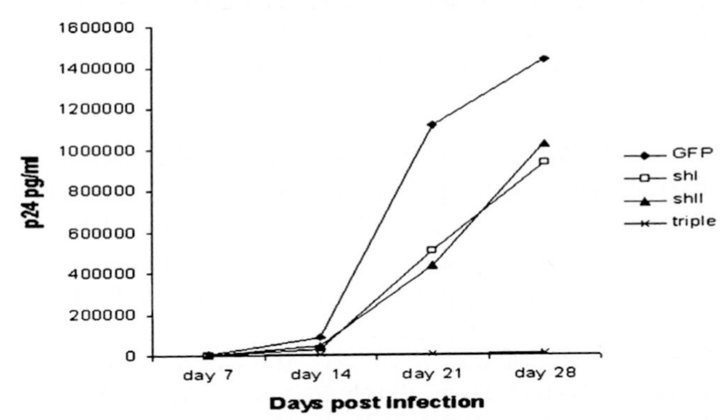

FIGURE 4. The triple construct provides long-term, potent repression of HIV replication in CD34$^+$-derived monocytes challenged with HIV-1 JRFL. Hematopoietic cells were isolated, transduced, and challenged with HIV as described previously.[19] The transduced constructs include single shRNAs targeting tat/rev or rev, the triple construct, or the parental backbone vector pHIV7. At the indicated times supernatant samples were harvested and monitored for HIV p24 levels.

targets spliced *tat/rev* transcripts but can also trigger degradation of unspliced genomic RNA transcripts.[18] The combined expression of these three RNAs provides complete protection against HIV infection (FIG. 4). We have also been able to demonstrate that the expression of the triple constructs in primary immune cells does not elicit any unwanted toxicities.[21] Taken together, the triple combination provides the most promising gene therapy approach for treating HIV infection.

WHERE DO WE GO FROM HERE?

All of the preclinical tests with the triple inhibitor-expressing lentiviral vector suggest that this approach is efficacious and safe for hematopoietic gene therapy. The vector has been modified to eliminate the EGFP marker gene, and has been produced under GMP conditions for use in a gene therapy trial. A clinical trial utilizing the vector is planned. Autologous HSCs will be used as the targets for transduction followed by infusion into patients in a bone marrow transplant setting. The trial will involve five AIDS/lymphoma patients. This group of patients has been shown to have long-term remission from their AIDS-related lymphoma as a consequence of autologous stem cell transplants.[26] Our goal is to incorporate gene therapy with our triple construct vector with the bone marrow transplantation to treat both their lymphoma and HIV infection. This will be the first clinical trial using a vector expressing multiple anti-HIV genes, and the first to use RNAi in a stem cell setting.

ACKNOWLEDGMENTS

This work was supported by grants AI29329, AI42552, and HL 07470 to JJR.

REFERENCES

1. RATHBUN, R.C., S.M. LOCKHART & J.R. STEPHENS. 2006. Current HIV treatment guidelines–an overview. Curr. Pharm. Des. **12:** 1045–1063.
2. LEVY, A.R. *et al.* 2006. The direct costs of HIV/AIDS care. Lancet Infect. Dis. **6:** 171–177.
3. JONES, R. & B. GAZZARD. 2006. HIV/AIDS pathogenesis and treatment options focusing on the viral entry inhibitors. Expert Rev. Anti Infect. Ther. **4:** 303–312.
4. MORRIS, A., H. MASUR & L. HUANG. 2006. Current issues in critical care of the human immunodeficiency virus-infected patient. Crit. Care Med. **34:** 42–49.
5. FANNING, G., R. AMADO & G. SYMONDS. 2003. Gene therapy for HIV/AIDS: the potential for a new therapeutic regimen. J. Gene Med. **5:** 645–653.

6. VAN GRIENSVEN, J., E. DE CLERCQ & Z. DEBYSER. 2005. Hematopoietic stem cell-based gene therapy against HIV infection: promises and caveats. AIDS Rev. **7:** 44–55.
7. MICHIENZI, A. *et al.* 2003. RNA-mediated inhibition of HIV in a gene therapy setting. Ann. N. Y. Acad. Sci. **1002:** 63–71.
8. CHEN, W. *et al.* 2000. Lentiviral vector transduction of hematopoietic stem cells that mediate long-term reconstitution of lethally irradiated mice. Stem Cells **18:** 352–359.
9. ENGEL, B.C. & D.B. KOHN. 1999. Stem cell directed gene therapy. Front. Biosci. **4:** e26–e33.
10. FOLLENZI, A., L.E. AILLES, S. BAKOVIC, *et al.* 2000. Gene transfer by lentiviral vectors is limited by nuclear translocation and rescued by HIV-1 pol sequences. Nat. Genet. **25:** 217–222.
11. GUENECHEA, G. *et al.* 2000. Transduction of human CD34+ CD38- bone marrow and cord blood-derived SCID-repopulating cells with third-generation lentiviral vectors. Mol. Ther. **1:** 566–573.
12. HANAZONO, Y., K. TERAO & K. OZAWA. 2001. Gene transfer into nonhuman primate hematopoietic stem cells: implications for gene therapy. Stem Cells **19:** 12–23.
13. MIYOSHI, H., K.A. SMITH, D.E. MOSIER, *et al.* 1999. Transduction of human CD34+ cells that mediate long-term engraftment of NOD/SCID mice by HIV vectors. Science **283:** 682–686.
14. SIRVEN, A. *et al.* 2000. The human immunodeficiency virus type-1 central DNA flap is a crucial determinant for lentiviral vector nuclear import and gene transduction of human hematopoietic stem cells. Blood **96:** 4103–4110.
15. TRONO, D. 2001. Lentiviral vectors for the genetic modification of hematopoietic stem cells. Ernst Schering Res Found Workshop, 19–28.
16. UCHIDA, N. *et al.* 1998. HIV, but not murine leukemia virus, vectors mediate high efficiency gene transfer into freshly isolated G0/G1 human hematopoietic stem cells. Proc. Natl. Acad. Sci. USA. **95:** 11939–11944.
17. MACGREGOR, R.R. 2001. Clinical protocol. A phase 1 open-label clinical trial of the safety and tolerability of single escalating doses of autologous CD4 T cells transduced with VRX496 in HIV-positive subjects. Hum. Gene Ther. **12:** 2028–2029.
18. LI, M.J. *et al.* 2003. Inhibition of HIV-1 infection by lentiviral vectors expressing Pol III-promoted anti-HIV RNAs. Mol. Ther. **8:** 196–206.
19. LI, M.J. *et al.* 2005. Long-term inhibition of HIV-1 infection in primary hematopoietic cells by lentiviral vector delivery of a triple combination of anti-HIV shRNA, anti-CCR5 ribozyme, and a nucleolar-localizing TAR decoy. Mol. Ther. **12:** 900–909.
20. LI, M. & J.J. ROSSI. 2005. Lentiviral vector delivery of siRNA and shRNA encoding genes into cultured and primary hematopoietic cells. Methods Mol. Biol. **309:** 261–272.
21. ROBBINS, M.A. *et al.* 2006. Stable expression of shRNAs in human CD34+ progenitor cells can avoid induction of interferon responses to siRNAs in vitro. Nat. Biotechnol. **24:** 566–571.
22. MICHIENZI, A., S. LI, J.A. ZAIA & J.J. ROSSI. 2002. A nucleolar TAR decoy inhibitor of HIV-1 replication. Proc. Natl. Acad. Sci. USA. **99:** 14047–14052.
23. CAGNON, L. & J.J. ROSSI. 2000. Downregulation of the CCR5 beta-chemokine receptor and inhibition of HIV-1 infection by stable VA1-ribozyme chimeric transcripts. Antisense Nucleic Acid Drug Dev. **10:** 251–261.

24. LEE, N.S. *et al*. 2002. Expression of small interfering RNAs targeted against HIV-1 rev transcripts in human cells. Nat. Biotechnol. **20:** 500–505.
25. SCHERER, L.J. *et al*. 2004. Rapid assessment of anti-HIV siRNA efficacy using PCR-derived Pol III shRNA cassettes. Mol. Ther. **10:** 597–603.
26. KRISHNAN, A. *et al*. 2005. Durable remissions with autologous stem cell transplantation for high-risk HIV-associated lymphomas. Blood **105:** 874–878.

Index of Contributors

Aartsma-Rus, A., 74–76
Adamis, A.P., 151–171
Allakhverdi, Z., 62–73
Allam, M., 62–73
Aronson, J.F., 116–119
Arzumanov, A.A., 103–115

Barrett, A.D.T., 116–119
Beasley, D.W.C., 116–119
Bergeron, D., 91–102
Brown, D.E., 103–115

Chattopadhyaya, J., 124–136
Cheung, J., 120–123
Contag, C.H., 52–55

Damha, M., 91–102, 124–136
De Winter, C.L., 74–76
Dietrich, I., 44–46
Dori, A., 77–90
Dunham, B.M., 47–51

Ebbinghaus, S.W., 27–30
Ferrari, N., 62–73, 91–102

Gait, M.J., 103–115
Gewirtz, A.M., 124–136
Golenbock, D.T., 31–43
Gorenstein, D.G., 116–119
Gruenert, D.C., 120–123
Guimond, A., 62–73

Hanai, K., 9–17
Heemskerk, J.A., 74–76
Herzog, N.K., 116–119
Hickerson, R.P., 56–61
Honma, K., 9–17
Huang, L., 1–8

Ilves, H., 52–55
Ivanova, G., 103–115
Ivins, B.E., 137–150

Janson, A.A.M., 74–76
Johnston, B.H., 52–55
Juliano, R.L., 18–26

Kalota, A., 124–136
Kaspar, R.L., 52–55, 56–61
Kaufmann, S.H.E., 44–46
Klinman, D.M., 137–150

Lamphier, M.S., 31–43
Landthaler, M., 56–61
Latz, E., 31–43
Leake, D., 52–55
Leary, J.F., 116–119
Lee, L.K., 47–51
Leube, R.E., 56–61
Lever, A.M.L., 103–115
Li, H., 172–179
Li, M., 172–179
Li, S.-D., 1–8
Li, Z., 47–51
Lomas, L.O., 116–119
Luxon, B.A., 116–119

Maeda, M., 9–17
Mangos, M.M., 91–102
Maurisse, R., 120–123
McLean, W.H.I., 56–61
Minakuchi, Y., 9–17

Nagahara, S., 9–17
Ng, E.W.M., 151–171

Ochiya, T., 9–17
Opalinska, J.B., 124–136

Paquet, L., 62–73, 91–102
Patzel, V., 44–46

Reigadas, S., 103–115
Renzi, P.M., 62–73, 91–102
Rossi, J., xi–xiii, 172–179
Roth, C.M., 47–51

SéGuin, R., 62–73
Sano, A., 9–17
Seyhan, A.A., 52–55
Sirois, C.M., 31–43
Smith, F.J.D., 56–61
Soreq, H., 77–90
Stankova, L., 27–30

Takeshita, F., 9–17
Tedeschi, A.-L., 91–102
Toulmé, J.-J., 103–115
Turner, J.J., 103–115
Tuschl, T., xi–xiii

Van Deutekom, J.C.T., 74–76
Van Ommen, G.-J.B., 74–76
Verma, A., 31–43
Vlassov, A.V., 52–55
Volk, D.E., 116–119

Wang, H., 116–119
Wang, Q., 52–55
Widdicombe, J., 120–123

Xie, H., 137–150

Yang, X., 116–119

Zemzoumi, K., 62–73
Zhao, X., 116–119
Zhilina, Z.V., 27–30
Ziemba, A.J., 27–30